Roy Boyd

TROPICAL FOREST CONSERVATION

TROPICAL FOREST
CONSERVATION

An Economic Assessment of the
Alternatives in Latin America

DOUGLAS SOUTHGATE

New York   Oxford

Oxford University Press

1998

Oxford University Press

Oxford   New York
Athens   Auckland   Bangkok   Bogota   Bombay
Buenos Aires   Calcutta   Cape Town   Dar es Salaam
Delhi   Florence   Hong Kong   Istanbul   Karachi
Kuala Lumpur   Madras   Madrid   Melbourne
Mexico City   Nairobi   Paris   Singapore
Taipei   Tokyo   Toronto   Warsaw

and associated companies in
Berlin   Ibadan

Copyright © 1998 by Oxford University Press, Inc.

Published by Oxford University Press, Inc.
198 Madison Avenue, New York, New York 10016

Oxford is a registered trademark of Oxford University Press

Library of Congress Cataloging-in-Publication Data
Southgate, Douglas DeWitt, 1952–
Tropical forest conservation : an economic assessment of the alternatives in
   Latin America / by Douglas Southgate.
       p.   cm.
   Includes bibliographical references and index.
   ISBN 0-19-510996-1
      1. Rain forest conservation—Latin America.   2. Deforestation—
Control—Latin America.   3. Forest ecology—Latin America.   4. Non-
timber forest resources—Latin America.   5. Sustainable forestry—
Latin America.   6. Habitat conservation—Latin America.
7. Ecotourism—Latin America.   I. Title.
SD414.L29S68   1998
333.75'16'0980913—dc21   97-34601

9 8 7 6 5 4 3 2 1

Printed in the United States of America
on acid free, recycled paper

Acknowledgments

This book has benefited from the advice and support offered by a number of colleagues and friends. First and foremost, William ("Jeff") Vaughn, of the Inter-American Development Bank, hired me, as a consultant, in 1995 to assess the contributions that nontimber extraction, low-impact logging, genetic prospecting, and ecotourism can make to tropical forest conservation in Latin America. The consultancy provided me with a unique opportunity to interview experts and to visit sites in Brazil, Costa Rica, and Ecuador. All this greatly enriched the content of chapters 4–7 of this book.

I deeply appreciate the comments on a draft text that I received from Vaughn and five other individuals: Bruce Aylward (of the Tropical Science Center in San José, Costa Rica), John Browder (of Virginia Polytechnic Institute and State University), David Simpson (of Resources for the Future), Steven Stone (of Cornell University), and Christopher Uhl (of Pennsylvania State University). Also very useful were reviews of my Inter-American Development Bank report provided by Glenn Prickett and John Reid (of Conservation International) and four of Vaugin's associates: Sergio Ardila, Arthur Darling, Kari Keipi, and Gil Nolet.

Among others, John Dixon (of the World Bank), Hans Gregersen (of the University of Minnesota), and David Pearce (of University College London) encouraged me to undertake an examination of tropical forest conservation measures being applied in Latin America. As work proceeded, Dixon and three of his World Bank colleagues—Luís Constantino, John Kellenberg, and Robert Schneider—furnished a great deal of useful information. So did Marc Dourojeanni, Michelle Lemay, and Raul Tuazón, all of the Inter-American Development Bank. Several other people deserve special thanks, including Alfredo Carrasco (of the Charles Darwin Foundation), Howard Clark, Douglas McMeekin, Roberto Ulloa, and Robert Vogel—all of Quito, Ecuador; Jaime Echeverría and Joe Tosi of the Tropical Science Center; Adalberto Veríssimo and his colleagues at the Instituto do Homem e Meio Ambiente da Amazonia, in Belém, Brazil; Jaime Acosta, Anabella Lardé de Palomo, and their associates at the Fundación Salvadoreña para el Desarrollo Económico y Social, in San Salvador, El Salvador; George Carrington Wood, of Haverford, Pennsylvania; Juan Carlos Quiroga of La Paz, Bolivia: William Possiel (of The Nature Conservancy); Clovis Scharappe-Borges and Vitória Müller of the Sociedade de Pesquisa em Vida Selvagem e Educaçao Ambiental, in Curitiba, Brazil; as well as Claudio González-Vega, Brent Sohngen, and other colleagues of mine at The Ohio State University. Two other people at Ohio State also need to be acknowledged: Chad Forster, my graduate assistant, spent many hours tracking down literature and other materials, and Janice DiCarolis prepared all the maps and line drawings.

Special thanks are due to Kirk Jensen and his colleagues at Oxford University Press, for the excellent advice and support given at every stage of the writing and publication process.

Of course, I am exclusively responsible for all the book's errors and omissions, and the views and opinions expressed herein are mine alone.

Finally, I am especially grateful to my wife, Myriam, and my children, Elizabeth and Richard, for being patient as I labored on this volume, which I dedicate to them.

Contents

Introduction

Responding to the Challenge of Habitat
Destruction in Latin America

Tropical deforestation has come to be an abiding international concern, and support remains strong around the world for arresting encroachment on threatened habitats in Latin America and other developing regions. Exactly how to accomplish this task, however, is still a subject of controversy.

Systems of national parks, like those that have been established in wealthy nations, have proven difficult to transplant to areas designated for special protection in Africa, Asia, and Latin America. More often than not, these areas are already inhabited, and local people, who tend to be quite poor, resent being told by outsiders that they must relinquish certain economic activities, move somewhere else, or both.

The difficulties of creating and maintaining an official protected area are illustrated by the case of Machalilla National Park, along the Ecuadorian coast. The region has been inhabited continuously for millennia and, when the reserve there was set up in the early 1970s, local property owners were promised payment for the land being taken from them. However, few of those individuals ever have received anything. This would have been a simple, though regrettable, instance of uncompensated confiscation had the government taken effective action to safeguard its formal claims on resources. But it did not. As of the early 1990s, only sixteen guards and other employees were assigned to Machalilla Park, which according to official maps takes in 467 square kilometers. Cattle grazing and fuelwood collection continue in virtually every accessible part of the reserve. Indeed, the argument could be made that, at least in Machalilla's case, public sector "protection" has accelerated environmental degradation since resource users, many of whom formerly had an ownership stake, now regard the area as a free, or open-access, resource.

By the middle 1980s, the need for an alternative approach to the protection of natural habitats in the developing world was becoming obvious. Rather than trying to suppress economic activity by local communities, those involved in the implementation of integrated conservation and development projects (ICDPs), as the new sort of initiative came to be called, sought to address those communities' desires for improved standards of living and to secure resource ownership, while at the same time meeting conservation objectives.

A typical ICDP has two main thrusts. The first is enhanced park protection, conventionally understood to include boundary demarcation, improvement of trails and facilities, and the training and deployment of guards. The second thrust is to enhance earnings in a surrounding buffer zone by promoting the sustainable harvesting of forest products and, quite often, nature-based tourism. The intention is for local people to give up economic activities, like agricultural land clearing, that result in substantial damage to natural resources in favor of more environmentally benign alternatives. Insofar as they do so, encroachment on parks should be diminished.

Although all this sounds attractive, very few ICDPs have enjoyed much success. A review of twenty-three such projects in Africa, Asia, and Latin America, for example, reveals that enlisting the necessary cooperation from local communities is often difficult (Wells and Brandon, 1993). In addition, ICDPs have been criticized on more general grounds. For one thing, ecotourism and other preferred activities are not always environmentally benign. Also, the roads and other improvements that are often needed for an ICDP to function simultaneously enhance the profitability of depletive lines of work. Southgate and Clark (1993) point out that, where labor is underemployed, local populations can adopt ICDP activities without giving up what they would do otherwise. For reasons such as these, observers like Dixon and Sherman (1990) have expressed doubts that parks and reserves can truly be kept intact by encouraging things like nature-based tourism or the sustainable harvesting of forest products in surrounding buffer zones.

Outline of the Book

That ICDPs and related initiatives have not enjoyed great success calls into question an approach to tropical forest conservation that has substantial prima facie appeal and has attracted considerable moral and tangible support—namely, the promotion of environmentally sound economic activities in or near natural habitats that are under threat.

A good way to assess the economic viability of this approach would be to draw on assessments of ICDPs, themselves. Sad to say, though, this is not possible. In one of the few contributions to the economic literature dealing with those projects, Simpson and Sedjo (1996) complain that just about all publicly available information can be described as self-laudatory, because agencies responsible for administration or funding are not keen to report disappointing results, because the projects involved do not have long track records, or both. Even the "gray literature," consisting of internal agency reports and evaluations, is not terribly informative, presumably for exactly the same reasons.

This book contains an economic assessment of forest conservation strategies predicated on the fostering of sustainable economic activities in and around threatened habitats. Little reference is made here to the ICDP literature, such as it is. Instead, activities often included in ICDPs are the subject of direct examination. Four case-study chapters, each devoted to a single type of activity, comprise the core of the book. The case studies yield insights, offered in the last two chapters, for a truly integrated strategy for ecosystem conservation and economic progress in Latin America.

A proper point of departure for a study of what has become a fashionable approach to the preservation of tropical forests and other threatened habitats is a survey of the extent and consequences of deforestation. Chapter 1 of this book contains such a survey. Rates of encroachment on tree-covered land vary, and there appears to have been a peak in the rates in the Brazilian Amazon during the middle 1980s. However, extensive tracts of land that have remained relatively untouched up to now are in jeopardy. If there were widespread conversion of forests into cropland and pasture land, the impacts on local climate and global biodiversity would be severe. Global warming could accelerate as well.

In chapter 2, various factors that contribute to land use change are analyzed. Among these factors are population and income growth, poorly articulated property rights, infrastructure development, the tendency of deforestation agents to neglect the environmental impacts of their activities, as well as direct and indirect subsidies for land clearing. Above all, the construction of new roads in previously inaccessible places sets off a cycle of ecosystem depletion that is all but irresistible for the actors directly involved.

The contribution that ICDPs and related initiatives might make to habitat conservation, in the face of all this, is addressed in the chapter 3. In particular, the commercial potential of extracting nontimber forest products and other sustainable activities is examined in conceptual terms, with special emphasis being placed on understanding the circumstances under which these activities actually benefit local populations.

Chapter 4 is about the harvesting of nontimber forest products. Experience in the Amazon Basin indicates that there are various impediments to that activity's economic and environmental success, including weak property rights, thin markets, and production outside of forest settings. In addition, a study of extraction of nontimber products in western Ecuador reveals a general tendency toward meager financial returns for the households that engage in harvesting. By contrast, processors and exporters tend to capture most of whatever profits are generated by the exploitation of nontimber resources.

Environmentally sound timber production is the focus of chapter 5. Investigation of various modes of timber harvesting and extraction in the eastern Amazon provides a clear picture of how logging evolves in frontier regions, and also yields the conclusion that sheer resource abundance discourages the sort of investment required for sustainable resource management. The latter conclusion is corroborated by experience gained in a sustainable forestry project carried out in the Peruvian Amazon with financial and technical support from the U.S. Agency for International Development (AID).

In chapter 6, there is a review of the empirical literature on the value of tropical forests as a source of raw material for biomedical research. Rudimentary estimates of that value, contained in early contributions to the literature, turn out to have been too high. The best available economic research suggests that the economic returns of genetic prospecting might be quite modest, particularly for forest dwellers. These returns are almost certainly too small to justify the investment in property institutions that is required to establish efficient markets for genetic information collected in the wild.

Nature-based tourism in Costa Rica and the Galápagos Islands is examined in chapter 7. Both places have drawn large numbers of international visitors in the 1980s and 1990s, and national economies have benefited enormously as a result. By and large, local communities, which are poorly equipped to provide the goods and services that ecotourists demand, gain little from the visits being made to nearby parks and reserves. In addition, there is a need to shore up the environmental base for tourism's continued success. To accomplish this task, entrance fees will have to be raised and other financing mechanisms will have to be exploited.

As demonstrated in this book, ecotourism, extraction of nontimber products, environmentally sound timber production, and genetic prospecting can, under the right circumstances, contribute to habitat conservation and improved living standards in selected areas. But in and of themselves, these activities cannot serve as a sound centerpiece for an integrated strategy for economic development and habitat conservation.

Much more can probably be accomplished by raising crop and livestock yields, so that agricultural land clearing is no longer needed to satisfy increasing commodity demands. Of even greater importance is human capital formation, which reduces the number of people for whom converting natural ecosystems into marginal farmland is an attractive employment option. Indeed, excessive encroachment on natural ecosystems in Latin America can be interpreted as an environmental manifestation of the more fundamental problem of rural poverty, resulting from inadequate human capital formation, and also from the lack of institutions—which economists call social capital—that are needed for markets to function robustly.

Available evidence, which is reviewed in chapters 8 and 9, suggests that a combination of sustainable agricultural intensification and human and social capital investment allows just about any country to raise material standards of living while keeping natural habitats intact. The prospects for Latin America's threatened ecosystems will be very bleak if accelerated development of the region's rural economy through productivity-enhancing investment does not occur.

1

Deforestation and Its Causes and the Challenge of Sustainable Forest-based Activities

1

Deforestation in the American Tropics

The Regional and Global Stakes

Latin America's natural habitats have been receding for centuries. Before the 1970s, though, nearly everyone in the region regarded the geographic expansion of agriculture and other sectors of the rural economy as an unmitigated benefit. Well into the twentieth century, primary forests and other relatively undisturbed ecosystems were extensive. Moreover, it was generally agreed that human domination of the landscape must increase as nations develop and economies grow. Hence, deforestation was equated with progress.

Economic assessments of widespread land use change occurring through the early 1960s do not exist, mainly because very few contemporary observers sensed that counting up benefits and costs and analyzing causes would be worth the trouble. Quite often, it must be said, a complacent view of deforestation was entirely justified, as a pair of examples illustrate.

First, clearing of agricultural land in the Guayas River Basin of western Ecuador began in the middle of the nineteenth century and continued through the early 1900s, mainly to accommodate a major expansion of cocoa production. After a lull between the First and Second World Wars, deforestation resumed in the 1940s, when banana plantations were first established (Bromley, 1981). With fertile soils and abundant rainfall, the watershed is where most of Ecuador's crops and livestock are grown today.

Second, southern Brazil has experienced dramatic land use change during the last few decades. Londrina, an urban center in the state of Paraná with a population of several hun-

3

dred thousand, had fewer than 20,000 inhabitants in the 1930s and was completely surrounded by forests. Coffee farming predominated between the 1940s, when agricultural land clearing began in earnest, and the middle 1970s, when growers abandoned the area in the wake of a devastating freeze. Soybeans soon became southern Brazil's major crop. Only a few small woodlots now remain in the interior of Paraná, which has become a center of oilseed production for the country and the entire world.

Nearly all concerned, including local residents, agree that more trees should be planted in places like the Guayas River Basin and southern Brazil. Likewise, it is widely conceded that soil erosion and chemical contamination must be reduced and that other environmental ills associated with agricultural development have to be curbed. However, no one doubts that deforesting land that is well suited to crop and livestock production has been, on balance, hugely beneficial. Land use change around, say, Londrina is not much more controversial than the migration of U.S. farmers into the Ohio River Valley would have been in the early 1800s.

The same cannot be said of the conversion of Amazonian forests into cropland and pasture, which a number of national governments began to encourage in the 1960s. Notwithstanding the enthusiasm that some officials and many settlers had for farming and ranching in the region, increased agricultural output was never the main purpose of colonization efforts. Instead, the primary goals were to ease demographic pressure in more densely populated regions and to solidify national territorial claims. General Emilio Garrastazu Medici, who headed Brazil's military government from November 1969 to March 1974, expressed both the objective and the approach succinctly when he spoke of "bringing men without land to a land without men" (Scartezini, 1985, p. 10).

The difficulties of large-scale agricultural colonization in Amazonia were apparent by the early 1970s. Nelson (1973) was among the first to point out the impediments to success in the humid tropics. Criticism was especially strong in regard to an ambitious program to settle 70,000 families along the Transamazon Highway, a road completed in 1972 that runs south of, and approximately parallel to, the Amazon River. Smith (1981) has found that only 3 percent of the land originally designated for colonization is suitable for agriculture. Furthermore, colonists suffered greatly from malaria because deforestation created breeding grounds for the

Anopheles mosquito (Morán, 1983). Accordingly, the program fell far short of planners' expectations; just 8,000 farmers and their families were living along the Transamazon Highway in 1980 (Smith, 1981).

Agricultural colonization, directed as well as spontaneous, has come under attack for being a threat to long-term forest dwellers. It is significant that Fernando Belaunde-Terry, who as Peru's president had championed settlement of the Amazonian rainforests until being deposed in a 1968 military coup, encountered staunch opposition from indigenous groups and their domestic and international allies when he tried to reinvigorate colonization projects after returning to the presidency, in 1980 (see chapter 5). Likewise, rubber tappers in the Brazilian Amazon have mounted a stiff resistance to cattle ranchers' encroachment on tree-covered lands (see chapter 4).

Of course, by the time that forest dwellers were starting to have some voice in discussions about the future of the places they live in and about the development of the resources that surround them, the adverse environmental consequences of land use change were starting to arouse substantial concern. Myers (1984) offered early warnings about the species extinction brought about by the alteration of tropical habitats. Also, deforestation appears to contribute to a rise in atmospheric concentrations of greenhouse gases, like carbon dioxide (Dixon et al., 1994).

The Magnitude of Land Use Change

Most likely, alarm over habitat loss in Latin America and other parts of the developing world peaked soon after 1987. That year's dry season was unusually pronounced in many parts of the Amazon Basin, which made it particularly easy to burn off natural vegetation. In addition, many Brazilians were eager to claim tree-covered land by putting it to an agricultural use because they sensed that impending constitutional reform would circumscribe that age-old tenurial convention.

With farmers and ranchers invading forests and reclearing fields that had reverted to bush, satellite cameras were able to detect an arc of flames extending south from Belém, a city near the mouth of the Amazon River, and west to lowland Bolivia. The consequences, it seemed, were global in scale and followed quickly. The summer of 1988 turned out to be one of the hottest that North America has ever experienced, and the claim that the Earth's atmosphere is growing warmer largely because of land use change close to the Equator was expressed in congressional hearings, newscasts, and the scholarly literature—and it acquired great credence. Many are now firmly convinced that deforestation is the leading cause of global warming.

As indicated in Table 1.1, few parts of the developing world experienced more rapid land use change than Latin America did during the 1980s. Expressed as a share of tree-covered land, deforestation was especially high in Mexico and Central America. But even in South America, which has half the world's tropical forests, the ratio of newly cleared land to remaining forests was comparable to what was being observed in most of Africa and southern Asia.

Natural habitats continue to be lost at a distressing rate in many parts of the American tropics. Myers (1988) has identified ten "hot spots" around the developing world, where biodiversity is threatened severely. In each location, farmers, ranchers, loggers, and miners are encroaching rapidly on forests that are home to large numbers of species, many of which are endemic to a small area. Two of the three most critical hot spots are in Latin America. In western Ecuador and along Brazil's Atlantic Coast, up to 90 percent of the original forest has been either seriously degraded or lost entirely.

Clearing seemed to abate somewhat in the Amazon Basin in the early 1990s. In Brazil, for example, deforestation was less than a third of the maximum rate observed during the middle 1980s, which might have been as high as 90,000 square kilometers per annum (World Resources Institute, 1990, p. 42). According to bulletins issued by the Brazilian Institute of Geography and Statistics (IBGE), annual deforestation between 1992 and 1994 amounted to approximately 15,000 square kilometers.

However, land use change appears to have accelerated since then. In 1995, the Brazilian Amazon had another pronounced dry season, of the sort that occurred in 1987, and there was a corresponding upswing in land clearing. According to one prominent environmental organization, annual deforestation there has risen 34 percent since 1992 (World Wildlife Fund, 1996). Likewise, the World Bank's Inspection Panel issued a report on the Planafloro Project, which is supposed to conserve natural habitats in the southern portion of the Brazilian Amazon, and the report contains an estimate of deforestation in Rondônia (a state that borders Bolivia): 4,500 square kilometers per annum (International Bank for Recon-

Table 1.1 Natural Forests in 1990 and Average Deforestation
Rates for 1981 through 1990 in Selected Countries and Regions

Country/Region	Area (thousand km²)	Annual Loss (%)
Central Africa	2,041	0.53
Tropical southern Africa	1,459	0.84
West Africa	556	0.96
South Asia	639	0.79
Southeast Asia	2,106	1.33
Mexico	486	1.22
Central America	195	1.85
Brazil	5,611	0.61
Andean region and Paraguay	2,418	0.93

SOURCE: Data from WRI (1996): 218–219.

struction and Development, 1997). If that figure is correct,
then the pace of land clearing throughout the Amazon Basin
is much higher than what IBGE reported it to be in the early
1990s.

The threat to natural habitats will not ease any time soon.
Even in the Brazilian Amazon, where most primary forests
have remained inaccessible and therefore intact up to now,
widespread human alteration of extensive areas could be im-
minent. Uhl et al. (1997) point out that wood harvested in
the region already comprises well over half of what is consumed throughout the country. Moreover, foreign interest in
Amazonian resources is growing, in part because of forest
depletion in southeast Asia. Within a few hundred kilometers of roads and navigable waterways, logging is either taking place or can be expected to occur early in the next
millennium. Where timber extraction is moderate to heavy,
as it tends to be in areas close to the highway that runs from
Belém south to Brasília, natural habitat recovery is often
slow. In addition, land made accessible by skidder tracks and
logging roads is frequently occupied by farmers and ranchers,
which further postpones the date when tree cover might be
reestablished (Uhl et al., 1997).

Uhl et al. (1997) do not project how much of the Amazon
Basin will be deforested by any particular date. But they do
provide a map indicating where timber harvesting, which is
becoming a primary initial catalyst for habitat loss, is hap-
pening or soon will be. Aside from the areas traversed by the
Belém-Brasília Highway and other major roads, which are

where roundwood production is concentrated, there is a wide swath of land south of the Amazon River where selective harvesting of mahogany (*Swietenia macrophylla* King) takes place. Furthermore, Uhl et al. (1997) anticipate a major expansion of riparian logging, both in *várzea* floodplains and farther into the forest, throughout the watershed.

If farmers and ranchers follow loggers into many of these areas, much of the Amazon Rain Forest will be lost.

The Impacts of Deforestation

If the conversion of forests and other natural habitats into cropland and pasture is bound to continue in the American tropics, then likely effects, both within the region and around the world, need to be considered.

One impact has to do with watershed services, defined to include reduced displacement and transport of sediments as well as stream-flow regulation. As Hamilton and King (1983) have stressed, the beneficial consequences of keeping a catchment area covered with trees (or the opposite effect, the watershed services lost because of deforestation) have been exaggerated at times. In the same vein, Mahmood (1987) claims that the reductions in reservoir sedimentation that result from forest conservation, tree planting, and related activities are often overestimated because the magnitude of "background" displacement and transport of soil and rocks, which take place regardless of what human beings happen to do, is not always fully appreciated.

Reviewing these and other contributions to the literature, Chomitz and Kumari (1996) arrive at three conclusions. First, deforestation has not been shown to be associated with large-scale flooding. Second, the removal of tree cover does not necessarily reduce water availability during the dry season in downstream areas; indeed, an opposite relationship sometimes holds. Third, the linkage between upstream land use and soil management and downstream sedimentation depends greatly on the specific characteristics of the drainage basin. In a large and relatively level basin, a long time is required for eroded soil and rock to reach a reservoir or a navigable waterway, where sediments have to be dredged (at a substantial expense), can interfere with the production of various goods and services, or both. In smaller, more steeply sloped watersheds, the damage is more immediate. However, the poor performance of, say, hydroelectric projects located in the latter sort of catchment frequently has as much to do with inadequate accommodation of background sedimentation as it does with inadequate control of erosive human activities, and it may have even more to do with the former.

Chomitz and Kumari (1996) also point out that the specific effects of deforestation on the local climate are difficult to determine. However, the evidence that those effects are substantial is mounting. In the Amazon Basin and other areas where primary humid forests are extensive, transpiration (i.e., the release of water into the air from trees and other plants) is the source of most cloud formation and precipitation (Salati and Vose, 1984). It follows, then, that removing vegetation causes local rainfall to diminish.

The possibility exists in some places for emergence of a mutually reenforcing cycle of deforestation and dehumidification. As the local climate grows drier, because of reduced tree cover, fire risks increase, due to the presence of dead vegetative matter that no longer is too wet and is therefore combustible. Although some trees can withstand a forest blaze, others cannot. In addition, fire destroys many of the seeds that lie on or in the ground, ready to sprout in the small openings that constantly are being created in the forest canopy as mature trees are toppled by the elements (Uhl and Kauffman, 1990). In other words, deforestation can lead to drier conditions, which can result in fires, which destroy forests and impede regeneration, and so on.

Climatic effects farther afield are the subject of much concern and debate. As mentioned above, for at least a decade, a causal linkage has been posited between deforestation and global warming. But this does not mean that the issue is entirely settled. For one thing, fossil fuel combustion, which occurs mainly in affluent parts of the globe as well as in China, Mexico, and other nations experiencing rapid industrialization, accounts for the lion's share of greenhouse gas emissions; tropical deforestation generally is reckoned to contribute less than 15 percent of the world total (Dixon et al., 1994).

Furthermore, it is not unanimously agreed that global warming is truly under way, at least at the pace that some have described. A commission of experts organized by the United Nations has examined simulations of global climate change obtained with the most advanced models available, and has concluded that there is a modest warming trend that cannot be ascribed entirely to natural causes (Houghton et al., 1996). However, a number of scientists dissent from this view. They point out that even the best global-climate models have serious shortcomings. Skeptics also contend that observed warming to date, which has been fairly modest, cannot be attributed to the greenhouse effect (Seitz, 1996).

Regardless of how controversies over global warming and deforestation's contributions to it are resolved, there is virtually no doubt that the loss of natural habitats near the Equator is the single most important cause of declining global biodiversity. Although they occupy little more than one-twentieth of the world's land area, tropical forests harbor a large share of all known plant and animal species, and many more that are not yet named (Myers, 1984; Wilson, 1988). Particularly in the hot spots that Myers (1988) has identified in Africa, Asia, and Latin America, much of this biological wealth is being extinguished at an alarming rate. This adds up to a dramatic and irreversible change in the nature of life on our planet.

Almost certainly, biodiversity conservation is the main reason to protect tropical habitats.

2

The Causes of Excessive Habitat Destruction

There seems to be no end in sight to tropical defor- estation, and the resulting environmental impacts are generally viewed with grave concern. Accordingly, quite a lot is being written about factors contributing to land use change.

Pearce and Brown (1994), who have compiled a compre- hensive review of the existing literature, have identified sev- eral lines of inquiry into the causes of deforestation in the developing world. Several investigators have examined the impacts of the macroeconomic turmoil that beset nearly every Latin American nation, and several countries in Africa and Asia, during the 1980s and early 1990s. Other research has to do with what Pearce and Brown (1994) refer to as com- petition for space. The excessive land use change that has resulted because farmers, ranchers, and other individual agents do not internalize deforestation's environmental costs, and because of inappropriate government policies, has been studied as well.

Macroeconomic Crisis and Competition for Space

Linkages between land use conversion and developing coun- tries' debt crisis and its aftermath are complex. On the one hand, incentives to mine natural resources, including for- ested ecosystems, so that debts could be serviced had to have been strengthened. On the other hand, fiscal austerity has put pressure on the public sector to diminish subsidies contrib- uting to deforestation. At the same time, less money has been available for building the roads needed to open up previously inaccessible regions for colonization. Accordingly, it is hardly surprising that some investigators are unable to discern any statistically significant impact of indebtedness. One exception is Capistrano (1994), who has found in a cross-national regression study that there has been a negative relationship at times between debt-service ratios (i.e., interest and principal payments divided by export revenues) and forest loss. She also reports that currency devaluation, which is usually an essential feature of structural adjustment and which increases the relative value of timber and other tradable commodities, appears to be a stimulus for deforestation.

Several statistical studies indicate that there is a positive connection between habitat destruction and population growth, population density, and rising incomes. Each of these three factors, Pearce and Brown (1994) point out, gives rise to increased human occupation of environmental niches.

The space competition hypothesis, as it might be called, does not square easily with some widely circulated views about the causes of deforestation. In particular, the finding that income growth and encroachment on natural habitats are related can be seen as contradicting the claim that poverty is an important contributing factor. The latter claim has been repeated often. Recently, for example, the Consultative Group for International Agricultural Research (CGIAR) reported that poor farmers in developing countries clear nearly 160,000 square kilometers of tree-covered land each year (Crossette, 1996).

Furthermore, the idea that demographic growth and rising standards of living lead inexorably to habitat destruction is hardly a reassuring one. Human fertility has fallen dramatically in many developing countries, because of rising female literacy, urbanization, and other factors. However, populations continue to increase since cohorts of women able to have children are large. The space competition hypothesis, interpreted simplistically, suggests that the only way to avoid additional loss of natural ecosystems in the face of population growth would be for Africans, Asians, and Latin Americans to grow poorer.

It is fortunate, then, that research bearing on the hypothesis raises at least as many questions as it resolves. Consider, for example, the fact that demographic increases in South America lead to about five times the deforestation that results as populations grow in Africa or Asia. In fact, an average of 0.75 hectares of forested land are lost when one person is added to South America's population; the comparable figures for Africa and Asia are 0.16 and 0.13 hectares per person, respectively. The reasons for the discrepancy between South America and the rest of the developing world are not clear. One might think that higher land-clearing coefficients for Brazil and neighboring countries are the result of forests being converted mainly into low-density cattle pastures. But the truth is that Asia is the place where pasture represents the more prevalent new land use, with 92 percent of the deforested area being dedicated to livestock production, as opposed to 47 percent in South America and 29 percent in Africa (Pearce, 1996).

Obviously, other things are going on, at least in the Western Hemisphere.

The Contribution of Market Failure

Another reason why deforestation is excessive in Latin America and other parts of the developing world is that individual agents of land use change do not consider all the costs of their activities.

The problem is analogous to the inefficiency that would arise if owners of steel mills were never liable for the soiled laundry, pulmonary illnesses, and other impacts suffered because of air pollution. In such a situation, steel prices would be too low and no market value at all would be placed on clean air. In response to these incentives, steel production and emissions would both be too high.

In the case of deforestation, farmers, ranchers, and other agents of land use change regard the disutility resulting from global warming and biodiversity loss as externalities. The difference between the land clearing that they opt for and the level where efficiency is achieved can be illustrated with a very simple diagram, of the kind that can be found in any standard text on environmental economics.

Depicted in Figure 2.1 are the marginal net private benefits of agricultural colonization, defined as the difference between the market value of crops and livestock produced on the last hectare cleared and the internal costs incurred on that same hectare. The latter would include clearing expenses, spending on fertilizer, vaccines, and other purchased inputs, and the opportunity cost of labor and management inputs. As a rule, marginal net benefits decline as deforestation increases, because progressively more remote or less productive land is being occupied, wages rise as more labor is employed, and so forth. Also shown in the same diagram are

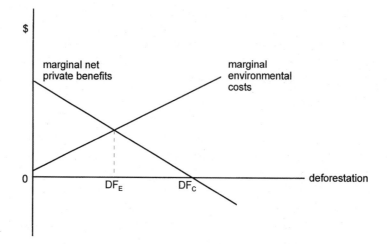

Figure 2.1 Excessive deforestation resulting from market failure.

the marginal environmental costs of deforestation, which are relatively modest if most land is left undisturbed, but which increase along with the clearing rate.

For deforestation to be efficient, a proper balance needs to be struck between net private benefits and environmental costs. In particular, agricultural colonization should occur only up to the level of DF_E, which is where marginal net benefits equal marginal environmental costs. But because farmers and ranchers neglect the latter, they will opt for a higher level—DF_C, which is where their marginal net benefits equal zero (Figure 2.1).

While the fundamental logic of the market failure explanation of excessive tropical deforestation is straightforward, quantifying the inefficiency depicted in Figure 2.1 is quite a challenge. The main reason is that deforestation's environmental costs, like most of the disutility resulting from air and water pollution, are difficult to estimate.

For example, Swanson (1995) argues that the world's ecosystems are becoming increasingly homogenized as an expanding and more affluent human population exercises progressively greater dominance of the landscape. At the same time, the values of relatively undisturbed habitats are rising because the diverse biological resources they contain might turn out to be useful in buffering the homogenized ecosystems (e.g., farms) on which people have come to depend so much, following some unforeseen shock. However, there is no monetary transaction corresponding to what society, as a whole, ought to be willing to pay for this service. Without such compensation, a marginal environmental cost curve, like that in Figure 2.1, cannot be plotted accurately, and quantification of excessive destruction of biologically rich habitats is imprecise.

There has been some empirical investigation of the values that people attach to keeping the developing world's biodiverse forests intact. Pearce (1996) has reviewed a series of debt-for-nature swaps effected in the late 1980s and early 1990s. In a typical transaction, an international conservation organization purchased the foreign debt issued by a poor country's government, usually at a substantial discount, and turned the debt over to the government in exchange for a certain amount of domestic currency to be made available for park protection and related initiatives. Dividing funding by the total area benefited, Pearce (1996) has found that the financial resources generated by nearly all the swaps amounted to $5 per hectare or less, and that the implicit per-hectare payment for quite a few was below $1. A similar analysis of the budget for the Global Environmental Facility (GEF), which the World Bank administers, yields implicit payments of the same order of magnitude (Pearce, 1996).

Research carried out by Kramer, Mercer, and Sharma (1996) suggests that the values that Pearce (1996) has calculated seriously understate what people really think the natural habitats in question are worth. Applying survey data to a contingent valuation model, Kramer, Mercer, and Sharma (1996) estimate that an average U.S. household would be willing to make a onetime payment of $29 to $51 to extend protection to an additional 5 percent of the world's tropical forests. Assuming that all affluent households place an equal value on the same outcome, Pearce (1996) concludes that the citizens of all the world's rich countries would offer $11 billion to $23 billion. Dividing these amounts by the area to be protected, which is 5 percent of approximately 20 million square kilometers around the world, yields $110 to $230 per hectare.

Obviously, these estimates of potential payments are much higher than the amounts of money that have actually changed hands through debt swaps and other arrangements. However, it is not clear that potential payments, were they to be made, would be large enough to cause a significant reduction in habitat loss. Schneider (1995) reports that real estate along Brazil's agricultural frontiers is normally bought and sold for $300 per hectare. Recently colonized parcels in northwestern Ecuador fetch as much as $1,000 per hectare, although the median value is below $500 per hectare. Since real estate prices generally reflect the present value of the future net returns that owners expect their holdings to generate, one doubts that, alone, biodiversity conservation payments of $110 to $230 per hectare would be enough to dissuade many farmers and ranchers from encroaching on tree-covered land.

An entirely different relationship appears to hold between real estate prices and payments made to sequester carbon in standing forests, so as to avoid changes in the world's climate. In an economic analysis of sea-level rise and other consequences that could result from global warming, Fankhauser and Pearce (1994) conclude that each ton of carbon emitted into the atmosphere causes around $20 in damages. That value can be applied to the emissions resulting from deforestation, which vary from approximately 100 tons per hectare, when closed, secondary forests in the tropics are converted to shifting cultivation or pasture, to more than 200 tons per hectare, when closed, primary tropical forests are cleared to make way for permanent agriculture (Brown and Pearce, 1994). The resulting cost estimates range from $2,000 to more than $4,000 tons per hectare.

Obviously, agricultural land clearing would be severely curtailed if farmers and ranchers had to pay $2,000 to $4,000 in global-warming damages for every hectare they cleared. Alternatively, landowners' interests in preserving tropical forests would be greatly enhanced if a way could be found to pay them for the carbon sequestered in undisturbed natural vegetation.

The potential for mutually beneficial economic exchanges between those in a position to keep carbon locked up in tropical forests and parties that are inclined or obliged to do something about global warming is intriguing. Members of the former group would be willing participants if payments received were at least equal to the income given up because farming or ranching is forsworn. Along Latin America's agricultural frontiers, that opportunity cost translates into a value of $5 to $10 per ton of sequestered carbon, which turns out to compare favorably with the expense of diminishing carbon emissions through other means. Payments to agents of tropical forest conservation also would be less than the taxes on emissions that some national governments, mainly in Europe, have considered (Fankhauser and Pearce, 1994).

It could be that an international market of sorts is developing for carbon sequestration services. In July 1996, for example, the Costa Rican and Norwegian governments announced that a prototypical agreement had been reached, under which the Norwegians will make $2 million available for reforestation and will receive, in return, bonds for the carbon stored in the planted trees (Oficina Costarricense de Implementación Conjunta, 1996).

However, carbon sequestration deals continue to be a novelty. One supposes that at least some participants are motivated mainly by a desire for favorable publicity. A true market for sequestration services will not emerge as long as carbon emissions remain lightly taxed and regulated, as they are around the world. That more serious measures to control emissions have not yet been adopted at the national level relates in part to doubts expressed in some influential quarters about whether global warming is really a serious threat (see chapter 1). But modest taxes and regulations might also reflect free riding. That is, some countries, hoping to benefit from climatic stabilization without paying for it, could be counting on other nations to raise domestic prices for fossil fuels and to purchase forest conservation rights. Behavior of this sort causes too little money to be raised for carbon sequestration. As a result, converting tropical forests into cropland and pasture continues to be inordinately attractive for rural people in the developing world.

The Role of Misguided Government Policies

Almost always, finding a way either to oblige people and firms to take their activities' environmental costs into account or to reward them for supplying beneficial environmental services is a considerable challenge. Building the political consensus needed for effective environmental legislation and regulations has taken many years in the United States and other affluent nations. A particularly vexing problem in the developing world is that governments' capacities to enforce any law, including something as basic as a traffic code, are modest in the extreme. Under the circumstances, many have concluded that attempting to tax or to regulate agents of tropical deforestation would not be particularly fruitful. A better alternative might be to focus on the reform of existing government policies that encourage the waste and misuse of timber resources, excessive agricultural land clearing, or both.

The rationale for the latter approach can be described, as Pearce and Brown (1994) have done, by slightly modifying the standard environmental economics model of inefficiencies induced by noninternalized costs (Figure 2.1). The new diagram (Figure 2.2) contains the same curve representing marginal environmental costs. But in addition to the marginal-net-private-benefits function in Figure 2.1, there is a new curve that indicates how those same benefits are augmented by direct or indirect subsidies from the government. As shown in Figure 2.2, deforestation is excessive for two reasons. The first, which is represented here and in Figure 2.1 by the gap between DF_E and DF_C, is market failure. The second, which economists call policy or intervention failure on the government's part, corresponds to the horizontal distance between DF_C and DF_S. The latter level is opted for by deforestation agents because that is where marginal net private benefits, including subsidies, equal zero.

There is no question that intervention failure has an impact. Repetto and Gillis (1988) contend that underpricing access to government-owned forests has encouraged private logging concessionaires to expand their activities beyond economically efficient levels. Also, some of the deforestation that has occurred in the Brazilian Amazon can be traced to subsidies of various sorts.

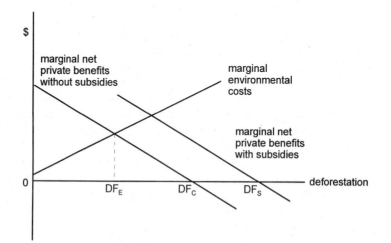

Figure 2.2 Excessive deforestation resulting from intervention and market failure.

Some of the fiscal inducements contributing to deforestation in Brazil have been indirect ones. For example, Binswanger (1989) has pointed out that the value which wealthy operators of large farms attach to rural real estate is inflated because agricultural income is taxed at a very low effective rate and because agricultural losses (calculated using liberal depreciation allowances and inflated estimates of operating costs) can be credited against nonagricultural income. Also, owning a large estate has tended to put one first in line to receive agricultural credit from the public sector on favorable terms. By contrast, most small operators pay little or no income tax, and are therefore not very interested in agriculture as a tax shelter. Also, their access to public sector credit is minimal, which means that their decisions about buying land are not influenced greatly by the prospect of getting a government loan that carries a low interest rate, that might be forgiven entirely, or both.

Policy-induced distortions of the sort Binswanger (1989) describes have contributed substantially to the concentration of Brazilian real estate in relatively few hands. Currently, in fact, 2.8 percent of the country's rural population owns 57 percent of the agricultural land. This in turn has precipitated violent clashes with the Movimento Sem Terra and other landless groups in recent years (Robinson, 1997). In addition, many of the small farmers who find it difficult to compete in the marketplace against wealthier individuals for prime farmland end up relocating to less promising areas, in the Amazon Basin, for example.

The competition for natural resources is unequal in many other parts of Latin America. For example, Heath and Binswanger (1996) blame the highly concentrated ownership of Colombia's best agricultural land and the confinement of small farmers to Andean hillsides and other marginal environments on agriculture's preferential tax treatment and policy-induced distortions in the financial sector.

Brazil is more unique in that, for a little more than two decades, the government made subsidies available for large-scale cattle ranching in the Amazon Basin. The system was put in place in 1966, when livestock projects approved by the Superintendency for Amazonian Development (SUDAM) first became eligible for tax credits. From 1971 through 1987, credits of nearly $3 billion (in constant 1990 dollars) were disbursed to the livestock sector, which also received almost $2 billion in loans carrying negative real interest rates (Schneider, 1995).

No doubt, subsidies greatly enhanced the financial attraction of livestock production in the Amazon for quite a few firms and individuals. Citing one enterprise-level study (Hecht, Norgaard, and Possio, n.d.), Mahar (1989) concludes that the typical SUDAM-approved project, which covered more than 20,000 hectares, would not have been profitable had tax credits not been available. Similarly, Browder (1988) has found that subsidies amounted to a large part of the accounting profits of many large cattle ranches.

However, it is important not to exaggerate the effects of direct fiscal incentives. Out of 84,000 square kilometers accounted for by all subsidized projects, 70,000 square kilometers were in just two states in the eastern Amazon, Pará and Mato Grosso. Of those 70,000 square kilometers, 20,000 square kilometers had been converted to pasture as of the middle 1980s. Since 156,000 square kilometers had been cleared in Pará and Mato Grosso during the preceding decade, no more than 13 percent of total deforestation in the two states took place on land owned by a person or firm receiving a subsidy (Schneider, 1995). If anything, the habitat loss caused by fiscal inducements accounted for less than 13 percent of the total since some of the 20,000 square kilometers would have been cleared even if no subsidies had been available.

Schneider (1995) offers another perspective on the limited impacts of direct subsidies, which the Brazilian government had phased out by 1990. He cites agricultural census data which show that, while the SUDAM program was still in effect, small- and medium-size ranchers were increasing their herds more rapidly than large operators, who were the program's main beneficiaries, were doing. Between 1980 and 1985, the number of cattle in Amazonian herds that had less than 50 head grew by 70 percent; for herds with 50 to 500 head, overall growth during the same period was 38 percent. Meanwhile, the number of cattle in herds larger than 500 head was just 17 percent higher for the region as a whole. Particularly interesting was the case of Rondônia, where the total number of cattle in herds larger than 200 head contracted by 2 percent between 1980 and 1985 and the total number in smaller herds rose by 80 percent during the same period, with practically no help from SUDAM.

Brazil is not the only place where direct subsidies' true contribution to agricultural land clearing is less than what some have believed. For example, Ledec (1992) has found that cheap credit for land clearing caused only 7-to-10 percent of cumulative deforestation in Panama through the 1980s. That share is considerably lower than the casual estimates which have been circulated at one time or another, and continue to be cited.

Factors Contributing to Ecosystem Mining

In terms of yielding useful answers to the question of what can be done to arrest encroachment on Latin America's natural habitats, much of the literature on the causes of tropical deforestation seems to lead to a dead end. Simply observing that either foreign debt, population growth, or poverty (or improved standards of living) makes a contribution yields few practical insights. Market failure certainly plays a role. But mechanisms for effecting international transfers to pay for carbon sequestration, biodiversity conservation, and other environmental services that tropical forests provide are at a nascent stage of development and are likely to remain so for many years. In addition, crop and livestock production in the Amazon Basin appears not to result primarily from the sort of financial inducements that were the subject of stern censure during the late 1980s.

Research that contributes most to an understanding of the factors driving deforestation, and which reveals most clearly what can be done to decelerate land use change, focuses on what happens when transportation infrastructure is developed in previously inaccessible areas. Schneider (1995) refers to this as road extensification, as opposed to road intensification. The latter involves making improvements in an accessible area's transportation network.

Needless to say, an important consequence of extended road networks is that it increases the availability of relatively cheap land in the rural economy. One response to this is accelerated migration from areas where real estate is relatively expensive to newly opened frontiers. For example, land prices in southern Brazil increased dramatically after the early 1970s, as a switch was being made in the region to large-scale, mechanized soybean farming. (Preferential tax treatment of agricultural income, credit subsidies, and other policies analyzed by Binswanger [1989] promoted the transition.) At the same time, new highways were being constructed in Rondônia and other parts of the Amazon. As the ratio of real estate values in Paraná and neighboring states to Amazonian land prices rose from two-to-one to ten-to-one or higher, small farmers in the former region found it worth their while to sell their holdings and to relocate to the frontier. Tens of thousands chose to do exactly that (Schneider, 1992).

A very similar situation arose in Guatemala during the 1970s. Due to improved economic prospects for sugar production, real estate prices were bid upward in the southern coastal lowlands. This in turn prompted ranchers to move their operations to areas in the northern part of the country that were just then becoming accessible (International Bank for Reconstruction and Development [IRBD], 1978, cited in Kaimowitz, 1996).

Once in a frontier setting, many migrants find it worth their while to engage in real estate speculation. For example, Colchester and Lohmann (1993) report that land price increases in northern Guatemala have consistently outstripped inflation since the 1970s. For at least part of this period, this trend reflected a long-term rise in international beef prices, which is important since Central America exported meat in significant quantities from the 1960s through the early 1980s and since most deforested land in the region has been dedicated to livestock production (Kaimowitz, 1996). In addition, real estate speculation is profitable where road networks are being put in place. For example, Ledec (1992) reports that infrastructure development in a part of Panama's Los Santos province caused local land prices to rise from $50 per hectare to $100.

Something else that almost always takes place in frontier regions, where renewable resources are relatively abundant, is "nutrient mining," as Schneider (1995) calls it. This process often commences with logging, which makes land more accessible in addition to yielding timber. Since previously uncultivated soils are relatively fertile, especially after vegetation has been burned, and free of pests, new colonists tend to spend a few years raising crops. But as nutrients are depleted and pests become more numerous, crop yields fall. At some point, it makes sense to take up extensive cattle ranching. If livestock yields fall below a certain level, the colonized parcel might be abandoned entirely.

Very important though it is, road extensification is by no means the only cause of ecosystem depletion along Latin America's agricultural frontiers. Most agents of depletion are poor people who have little or no access to formal financial markets; any credit they receive from informal sources tends to carry very high real interest rates (Schneider, 1995). At these high rates, depleting environmental wealth is much more remunerative than investing in land improvements is.

In addition, frontier tenurial arrangements are consistent with, and often promote, nutrient mining. Where there are no formal property rights, informal agricultural use rights are the norm. Furthermore, the latter regime was for many years codified in colonization laws, which were enforced by agrarian reform agencies. For example, Macdonald (1981) reports that, during the 1970s, the indigenous community of Pasu Urcu, in the Ecuadorian Amazon, abandoned swidden agriculture, which involves managing the successive growth of a wide variety of useful plants as well as periodic fallowing, and converted forests to pasture. They did this because government officials informed them that, if the change were not made, their land would be declared "idle" and therefore open to occupation by agricultural colonists who were 50 kilometers away at the time.

Settlers moving in large numbers to Latin American rain forests have felt no less obliged to clear land than Amerindians had, in order to safeguard formal or informal tenure. As Mahar (1989) points out, the "improvements" that a Brazilian colonist must make on his or her holding in order for the state to recognize his or her property rights have long been equated with deforestation.

It cannot be denied that ecosystem depletion involves a substantial drawing down of environmental wealth. This does not mean, though, that converting forests into cropland and pasture is inherently inviable. It would not make much sense, for example, to spend $300 per hectare on pasture regeneration that would lead to a tripling of net returns (not counting investment expenditures) if the same outcome could be obtained by purchasing more land at a per-hectare price of $150 or less. Instead, allowing yields to fall and then relocating to a new site is an economically rational response to land abundance: "With accessible land sufficiently cheap, it is more profitable to move the farm to the nutrient-rich, pest-free environment, than to import the fertilizers and pesticides to the farm" (Schneider, 1995, p. 16).

The cycle of depletion, abandonment, and relocation is given new life every time a new road is constructed through a virgin territory. It ceases only after all accessible land has been colonized, by which point real estate prices are rising, in response to increased resource scarcity, and productivity-enhancing investment is starting to be financially attractive.

3

Putting an End to
Ecosystem Depletion

Beginning in the 1960s, infrastructure development created myriad opportunities to exploit environmental wealth in the American tropics. For example, there were only 300 kilometers of paved roads in the Brazilian Amazon in 1960 (Mahar, 1989). In 1964, a federal highway traversing the eastern fringes of the watershed was completed to link Brasília and Belém. During the next dozen years, as many as 320,000 people migrated to the road's zone of influence (Katzman, 1977), where conditions for crop and livestock production are, for the most part, better than conditions found in the watershed's interior.

Similar events have unfolded in southern and western Amazonia and in other parts of the Western Hemisphere. The population of Rondônia, which has a larger area than many European nations have, was just 70,000 in 1960, and the outside world could be reached from there only by boat or by a small airplane. Since then, a federal highway and a number of feeder roads have been built, tens of thousands of migrants have arrived, and statehood has been achieved. Cumulative deforestation was approaching 25 percent in the late 1980s, and has surpassed that level since then (Mahar, 1989). Northeastern Ecuador was an inaccessible backwater until the late 1960s, when commercial petroleum deposits were discovered there. The infrastructure development that ensued has led since then to migration and agricultural land clearing on a large scale (Bromley, 1981). Likewise, the extensive deforestation that Central America has experienced has much to do with the construction of all-weather roads in the region;

overall highway length increased from 8,350 kilometers in 1953 to 26,700 kilometers in 1978 (Williams, 1986).

The pace of road extensification in Latin America diminished considerably after the middle 1980s. Fiscal austerity, brought on by the debt crisis, constrained investment in public works. Also, the World Bank and the Inter-American Development Bank (IDB) grew reluctant to support projects that might harm the environment, in large part because of pressure brought by supporters of indigenous rights and by international conservation organizations.

Private banks are assuming a much larger role in the financing of infrastructure development in Latin America. The corresponding decline in the importance of multilateral lending institutions is a consequence of the length of time they usually take to design a project, to carry out social and environmental reviews, and to win approval from boards of directors and host governments. Regardless, it is entirely possible that roads and other high-impact projects will be built without any multilateral bank involvement or environmental review whatsoever.

A case in point is the Pangue Dam, on Chile's Bío-Bío River, which was originally supported by the International Finance Corporation (IFC). According to press accounts, a confidential report commissioned by the president of the World Bank, with which the IFC is affiliated, was critical about the lack of enforcement of environmental conditions. When an attempt was made to rectify this, Chile's dam-building agency responded by paying off the IFC loan in full and borrowing from Dresdner Bank. Not only did the private source impose fewer environmental restrictions, but it charged a lower interest rate as well (*Economist*, 1997a).

Those who lobby in Washington, D.C., where the World Bank and the IDB are located, should keep this experience, which is hardly unique, in mind as they decide how best to attach environmental conditions to infrastructure development in Latin America. One thing that is perhaps working in their favor where highway construction is concerned is that many of the areas which remain inaccessible do not have the sort of agricultural potential needed to justify major expenditures on highways and related improvements. Investing in such areas should hardly be appealing to a bank or any other institution interested in maximizing profits. For this reason, road extensification is not expected to accelerate dramatically in most Latin American hinterlands, at least for the time being.

The Articulation of Formal Land Tenure

As frontiers close, the attractions of land improvement grow relative to those of ecosystem mining. This, in turn, causes property rights to evolve where colonization already has taken place. That is, increased resource scarcity induces the development of formal tenurial institutions, in which government assumes responsibility for protecting landowners from trespassers. In more remote areas, where formal property rights do not yet exist, landowners must fend for themselves, sometimes threatening interlopers with violence.

Progress toward stronger property rights is not always steady. In particular, people still draw on traditions of agricultural use rights to justify their attempts to acquire land for free, even after changes have been made in agrarian reform legislation and other legal arrangements that justify this sort of action. In August 1996, for example, there was a take-over of several hundred hectares in northwestern Ecuador owned by the country's leading wood products firm, which with full governmental approval and encouragement has been applying various measures to limit environmental damages and to enhance future timber production at the site. It is safe to assume that the participants felt that they could take advantage of the institutional flux that normally accompanies a change in the national government, of the sort taking place at the time.

Incidents such as these, and there have been many, demonstrate that the rule of law is not as strong as it should be in the Latin American countryside. Nonagricultural uses of land, like forestry, are greatly discouraged where agricultural use rights are the sine qua non of traditional tenurial arrangements. Even intensified crop or livestock production is discouraged since the effort devoted to raising productivity could instead be used to acquire more land, by clearing it of course.

Nevertheless, the more general trend is toward a property rights regime favoring investment, which Alston, Libecap, and Schneider (1996) have documented for the case of Pará. Furthermore, investment is indeed starting to occur where formal property institutions have been established. Recent trends in the livestock industry in a part of the state southeast of Belém, where livestock production got under way in the 1960s, are illustrative in this regard.

Mattos and Uhl (1994) have found that most pastures in the area are not being allowed to deteriorate over time as a prelude to abandonment and relocation. Instead, medium- and large-scale cattle producers have been rehabilitating their holdings, spending $260 per hectare, on average, on plowing, fertilizing, and reseeding; these measures raise annual live-weight production from 45-to-65 kilograms per hectare to 150-to-250 kilograms per hectare. Given the prices prevailing around 1990, the return on investment is 13 percent to 14 percent per annum. On a per-hectare basis, spending on pasture regeneration by smaller operators, who tend to specialize in calf and dairy production, is slightly lower; but the annual return on their investment is a little higher, 16 percent.

What Happens to Pioneer Nutrient Miners?

Cattle ranchers, many of whom operate on a large scale, have not been the only beneficiaries of agricultural settlement in the American tropics. A study of 440 land settlement projects undertaken by Brazil's National Institute for Colonization and Agrarian Reform (INCRA) in every part of the country reveals that the Amazonian projects have enjoyed some economic success. To be sure, the participating households have endured hardships, as indicated by high rates of infant mortality; in addition, they move from place to place more often than do farmers in the rest of Brazil (FAO/UNDP/MARA, 1992, cited in Schneider, 1995). But at the same time, the financial rewards are, for them, relatively good. Learning about crop and livestock production in a new environment can, and often does, require a lot of time and trial and error (Morán, 1989). However, the beneficiaries of INCRA's Amazonian program have enjoyed higher incomes and rates of asset accumulation than have their counterparts in all other parts of the country, with the exception of the prosperous southern states (FAO/UNDP/MARA, 1992, cited in Schneider, 1995, p. 12).

It is of paramount importance, though, to keep in mind an observation of Morán's (1989), which is that the people who thrive once property rights have grown stronger and human capital, communications infrastructure, and improved transportation networks have all begun to accumulate are usually quite different from those who initially colonize the frontier. To be specific, the beneficiaries of institutional and economic development tend to be better educated and often live in urban areas; they also have reasonably good access to governmental agencies and formal financial markets. By contrast, pioneer nutrient miners are poorly educated and they are apt to find that dealing with an agency, bank, or any other institution is a frustrating experience.

Where formal property rights have not been established, the advantages of people who are more affluent and well connected count for little. Also, the opportunity costs of the time they devote to defending a claim in a remote region, where the government's presence is negligible, are not covered by that activity's returns. However, those same returns are sufficiently attractive to colonists, whose time carries a low opportunity cost.

The tables turn once government begins to guarantee land tenure. Landowners are no longer obliged to occupy their holdings at all times so as to keep trespassers in check. In addition, the articulation of property arrangements makes it possible for formal financial institutions to function. When this happens, people with good access to those institutions—which lend money at interest rates substantially below what informal lenders charge—can borrow money to buy real estate, quite often from the original occupants. Wealthier people with better connections have other advantages. Their management skills tend to be superior. Furthermore, they find it relatively easy to arrange for government to provide them with land protection services.

Morán (1989) has documented that institutional and economic development in a frontier region is accompanied by a transition from one type of owner to another. Also, the survey of INCRA land settlement projects reveals that turnover of holdings has been high in the Amazon Basin (FAO/UNDP/MARA, 1992, cited in Schneider, 1995, p. 12), which is consistent with the transition in land ownership.

After selling out or otherwise giving ground in areas undergoing development, pioneer nutrient miners' choices are very limited. Indeed, those choices are not appreciably different from those facing the rural poor throughout Latin America. If they do not migrate, their rudimentary skills permit them to hold only menial, low-paying jobs. Seasonal farm labor and casual employment in the wood products sector would be two examples. Another choice, which does not exclude the first, is to support oneself and one's family as best as one can on a small holding, located perhaps in an area where property rights remain weak or nonexistent.

Another option is to abandon the countryside, and then to crowd into the urban slums where hundreds of thousands of people already live in Belém, Manaus, and a number of medium-size cities. As Browder and Godfrey (1997) indicate, this is the most frequently exercised option, which explains why the urban population of the Brazilian Amazon amounted to nearly 61 percent of the region's total in 1990 and continues to grow.

Saving Tropical Forests while Helping the Rural Poor

If institutional and economic development did not displace pioneer nutrient miners, the fate of Latin America's natural habitats would be determined in a relatively straightforward manner by economic incentives. That is, individuals with formal property rights could be counted on to behave no differently than, say, their counterparts in rural areas of the United States. With the state guaranteeing transferable and exclusive rights for resources, no discount would be applied to forestry earnings, reflecting dispossession risks associated with nonagricultural uses of land. Instead, landowners' choices would proceed from a direct comparison of future discounted net returns for various land use options. If, for example, the present value of earnings from forestry exceeded the present value of ranching income, then land would be dedicated to timber production. Likewise, a hike in real interest rates would lessen the relative appeal of land use options, like forestry, that yield revenues only after many years have passed.

Furthermore, manipulating land use in places where formal property rights are respected would involve little more than the imposition or withdrawal of taxes or subsidies. For example, more tree-covered land would be maintained if carbon sequestration payments were offered and if landowners found that the discounted sum of those payments and forestry income exceeded the present value of ranching income. Another way to arrest deforestation would be to remove whatever subsidies may exist for crop or livestock production.

However, the displacement of pioneer colonists that results from institutional and economic development means that habitat protection requires much more than just getting property arrangements and prices "right." Many, if not most, poor people from the countryside who do not move to a major city can be counted among what might be called a reserve population of potential nutrient miners, which stands ready to take advantage of any opportunity to engage in ecosystem depletion resulting from renewed road extensification, from appreciating values of open-access forest resources, or both.

It is widely accepted that habitats will never be safe as long as the rural poor are neglected. Indeed, there is a general sense that they must be full participants in the transition from nutrient mining to ecosystem management that must take place if natural habitats are to remain intact in places like the Amazon Basin. This sentiment is reflected in policies adopted recently by multilateral lending institutions. For example, the World Bank has stated that it will support forestry only if it is sustainable and only where there has been a "clear definition of the roles and rights of . . . forest dwellers" (IBRD, 1991, p. 66). Likewise, the IDB has committed itself to taking advantage of "opportunities to aid in the conservation of biological diversity" while also making sure that local communities share in the "benefits of sustainable forest management" (IDB, 1994, p. 34).

Nongovernmental environmental organizations also recognize the need to reconcile habitat protection with the enhanced well-being of local people. One such organization, which is active throughout the Western Hemisphere, is The Nature Conservancy (TNC). Its president, John Sawhill, has stated that he and his colleagues are "concentrating more on strategies that address . . . *the* conservation issue of the 1990s: integrating economic growth with environmental protection" (Howard and Magretta, 1995, p. 111).

There are various ways to protect the environment while at the same time helping impoverished local communities. One approach, which is examined more closely in chapter 9 of this book, is to improve education and public health services so that individuals who otherwise would engage in nutrient mining can instead compete successfully for jobs that pay better and at the same time create less resource depletion.

Human capital formation has been a feature of many of the habitat conservation projects that development agencies, multilateral lending institutions, and international environmental organizations have financed and/or implemented. But at least as much emphasis has been placed on encouraging individuals and groups living in or near threatened habitats to take up environmentally benign economic activities. These include the harvesting of nontimber products, logging in ways that do little damage to soils and remaining vegetation, collecting biological materials used in pharmaceutical and other research, and nature-based tourism.

Each of these four activities is analyzed in the next four chapters. Except under special circumstances, none turns out to be especially remunerative for local communities. One reason for this is that opting for something like sustainable forestry is unlikely to lead to an appreciable increase in the wages received by unskilled local labor. Another reason is that just about all the natural resources that environmentally benign economic activities draw on are abundant. As a rule, timber and other forest resources are not economically scarce. Likewise, there is no general shortage of ecotourism sites, although it is true enough that a few travel destinations, like the Galápagos Islands, are highly unique and desirable places. Except where a local community is fortunate enough to control access to an unusually valuable resource, local well-being will not be affected very much by the promotion of sustainable alternatives to ecosystem mining.

The challenge to a strategy for habitat conservation and local development that is predicated on the sustainable exploitation of renewable forest resources can be put in proper perspective by applying a conceptual framework originally developed by von Thünen (1866). The exceptional case where the strategy makes a great deal of economic sense will be considered first. After that, the more typical case, in which sustainable forestry is not very appealing, can be examined.

Opting for Sustainable Forestry Close to a Marketing Center

Even though, as already mentioned, aggregate resource stocks are more than ample, individual holdings can still be valuable because they are unusually productive. This possibility is not examined here. Instead, the focus is on another determinant of scarcity, which has to do with location. In particular, it can be shown that forests located close to a market can have value if transporting the products they yield is expensive and if no competing land use is more profitable.

If moving a given quantity of output a certain distance is indeed expensive, locational values are attached to advantageously situated resources. These values can be illustrated in the context of a very simple rain forest economy, one with a single commodity produced at a constant average cost, C. With the expense of transporting a unit of output one kilometer being defined as T, marginal cost, MC, is a simple linear function of distance from the market, D:

$$MC = C + T \times D. \tag{3.1}$$

Spatial equilibrium in the rain forest economy with one good is depicted in Figure 3.1A. At the extensive margin, D^*, the price that clears the market for the jungle commodity, P^*, equals per-unit production costs plus the expense of delivering output to buyers, nothing more and nothing less:

$$P^* = C + T \times D^*. \tag{3.2}$$

Beyond the extensive margin, combined costs exceed P^*, which means that production will not take place. But at inframarginal locations (i.e., those within D^*), sales revenues exceed the sum of production and transportation expenses. In other words, a rental premium, comprising the difference between the market-clearing price, P^*, and $C + T \times D$, is received for every unit of output.

If T is sizable, which would be the case if the jungle commodity were perishable, bulky, or both, then the MC curve

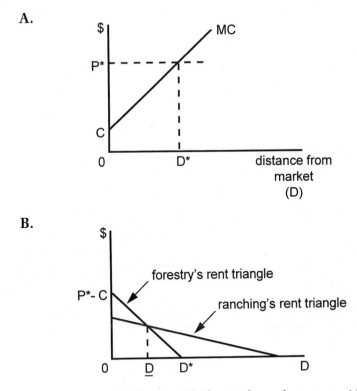

A.

B.

Figure 3.1 Spatial equilibrium. **(A)** The single-product case, with high per-unit transportation costs; **(B)** Land use succession for the two-sector case.

will be steeply sloped, as illustrated in Figure 3.1A. Producers located close to the market will collect sizable per-unit rents. Moreover, those rents will be directly manifested in resource values, since prices of inframarginal real estate will be bid upward in direct proportion to the gap between P^* and the costs of production and transportation. For example, what someone would pay for one year's use of an inframarginal hectare of forest yielding Q units of output would be:

$$\text{yearly land rents} = (P^* - C - TD)\, Q. \qquad (3.3)$$

With T being large, resource value declines precipitously as one moves away from the market. Yearly land rents at the extensive margin are nil. Beyond D^*, resources would be occupied only if a subsidy were offered, to cover the negative difference between P^* and $C + TD$.

To complete the analysis of forests' locational values (or, to be more precise, the economic circumstances under which property owners decide to maintain tree cover), one must reckon with the possibility of competing land uses. In particular, a series of land uses, not just the declining value of a single use, observed at progressively less advantageous locations must be identified. For simplicity's sake, suppose that ranching, which requires the clearing away of natural vegetation, is the only alternative to harvesting the jungle commodity. We are interested, then, in determining the distance from the market at which there is a succession in land use from forestry to livestock production.

To find the succession point, D, a rent triangle, which indicates yearly resource values (defined in equation 3.3 for the case of forestry) at varying distances from the market, must be plotted out for each of the two sectors. One corner of the triangle for forestry is found at D^* on the horizontal axis (see Figure 3.1B). The height of the same triangle corresponds to the rents captured by the most centrally located producer, who, by definition, incurs no transportation costs:

$$(P^* - C)\, Q. \qquad (3.4)$$

A rent triangle for ranching is plotted as well in Figure 3.1B. Its gentler slope is a signal that per-unit transportation costs in the livestock sector are lower than what they are in the market for the jungle commodity.

By definition, ranching and forestry are equally profitable at the succession point. Beyond D, livestock production represents the dominant land use, since it yields higher resource rents than forestry does. At lesser distances, property owners opt for production of the jungle commodity, and forests are left standing. (They also remain intact beyond ranching's extensive margin, which occurs where the livestock sector's rent triangle intersects the horizontal axis.)

Since time and effort are needed to delineate property rights, formal tenure will be established first for inframarginal land dedicated to uses that have a relatively high value. In the two-sector model, with C_{PR} standing for the annualized per-hectare cost of identifying and enforcing property rights, the frontier of institutional development occurs D_{PR} kilometers from the market. Beyond that frontier, claims on resources, which are worth less than C_{PR}, are not guaranteed by the state. At less remote locations, property rights are respected, regardless of whether land is used to raise cattle or whether forests are conserved, because annual returns to land exceed C_{PR}.

Forestry's Locational Values When Transportation Costs Are Low

It has to be repeated at this point that Figure 3.1's depiction of locational values and land use succession is based on the assumption that moving forest commodities from place to place is expensive, which implies that the MC curve in Figure 3.1A and the hypotenuse of the forestry rent triangle in Figure 3.1B are both steeply sloped.

To be sure, transportation costs for some commodities extracted from Latin America's tropical forests are high. A few jungle fruits, like the açaí (*Euterpe oleracea* Mart.) harvested near Belém (see Chapter 4), appear to be cases in point. But most forest commodities are no more difficult to move around than other goods are, and quite a few are relatively easy to transport. Neither timber nor latex, for example, is as perishable or bulky as most crops are. The implication of this is that hypotenuses of rent triangles for various parts of the forestry sector have much gentler slopes than what is depicted in Figure 3.1.B.

In addition, many forestry rent triangles are of modest height. This frequently occurs because the prices of close substitutes for forest commodities (e.g., timber or latex harvested on plantations) are low, which prevents market prices for those forest commodities from rising very high. Also, production costs in forest settings are often high, because of limited infrastructure, difficult terrain, and so forth. Something else that keeps forestry rents in check is the mobility of other factors of production. For example, small mills for cutting logs into boards are not very difficult to move from one site to another. If stumpage values (i.e., the prices paid for standing timber) rise in one place, there is not much to stop those mills from being shifted to locations where raw materials remain cheap. For all these reasons, forestry tends not to be a land use characterized by sizable locational rents.

What all this means is that, different from the situation depicted in Figure 3.1B, tree cover is more likely to be maintained outside, rather than within, the frontier of property rights development. Indeed, the succession from agricultural land use to forestry often occurs far beyond that frontier. As illustrated in Figure 3.2, the locational rents captured by nutrient miners exceed the locational value of forestry for the considerable distance lying between D_{PR} and the succession point, D. This implies that harvesting of forest products is confined to the most remote reaches of a rural economy.

Returns for Forest Dwellers' Labor

Where nature is bounteous, or transport costs are low, or factors of production other than environmental inputs are mobile, or there is some combination of the three, forest dwellers do not stand to gain very much by controlling access to the natural resources that surround them. It is not inconceivable, though, for those people to receive high incomes in the forestry sector by being paid a lot for their work and expertise.

Of course, wages and salaries in any labor market are determined by the interplay of demand, which is a function of workers' productivity and the price of whatever is being produced, and labor availability. Where output prices are low,

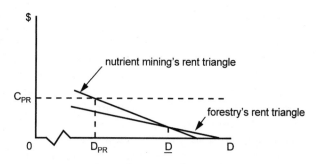

Figure 3.2 The succession from nutrient mining to forestry beyond the frontier of property rights development.

high earnings can occur only if productivity is high and the supply of labor is inelastic.

All available evidence suggests that the productivity of labor employed in logging, extraction of nontimber products, and other resource-based enterprises in Latin America's tropical forests is low. Biodiversity, itself, can be a cause of modest returns for labor. In forests featuring a great variety of species, for example, it is usually the case that useful organisms are widely dispersed, which implies that workers spend much of their time looking for something, as opposed to extracting it.

At the same time, the supply of labor tends to be elastic in places like the Amazon Basin, mainly because workers' mobility across regions and economic sectors is high. Romanoff (1981, cited in Browder, 1992b) has found, for example, that it is common for Amazonian rubber-tapper households to migrate every few years in the hope of gaining some sort of improvement in their meager earnings. Even when relocation does not occur, rural labor can move into and out of the cash economy (i.e., out of and into subsistence farming, hunting, and gathering) with ease.

When labor supply is highly elastic, wages cannot be expected to rise much above subsistence levels, even if major productivity increases occur in logging, extraction of nontimber products, and other forest-based occupations. Nor is the demand growth resulting from the opening up of new lines of work, in ecotourism or genetic prospecting, for example, likely to lead to substantially higher earnings for those living in or close to tropical forests.

The economic agents who are in the best position to benefit if launching or expanding an enterprise in a remote location proves to be remunerative are those furnishing factors of production that have low, as opposed to high, supply elasticities. Management, organizational, and marketing expertise is much more apt to fall into this category than are natural resources and unskilled labor. In addition, the firms that provide this expertise often face limited local competition, and sometimes none at all. No doubt, this enhances their ability to capture whatever profits are generated by activities like nature-based tourism and genetic prospecting.

Distributional Concerns and Human Capital Formation

By no means would a finding that local people gain little or nothing from an activity yielding positive environmental spillovers, like diminished pressure on natural habitats, mean that the activity should not be pursued. If the activity's benefits, including what people in other parts of the country or the world are willing to pay for spillovers, exceed its costs, then it satisfies the standard criterion of economic efficiency, and is worth pursuing.

Even if implementation of an efficient project benefits mainly people who are relatively affluent, incorporating re-distributional measures into the project is not warranted, if mechanisms, like a progressive income tax, are already in place to reduce disparities between the rich and the poor.

Transfers to impoverished local communities are an important component of project design, though, if they are needed to win the behavioral changes required to secure environmental benefits. For example, providing tree seedlings free of charge might be a good way to accelerate the reforestation of upper watersheds. Likewise, furnishing financial and technical assistance for non-timber products extraction, genetic prospecting, and so forth might diminish encroachment on rain forests and other species-rich tropical habitats.

It hardly needs to be said that the opportunity costs of effort dedicated to saving threatened ecosystems in the developing world need to be given close attention. This is especially true when the central thrust of a project is to build up local human capital that is specifically tailored to environmentally sustainable commercial activities. For example, investing in the specific skills that forest dwellers need in order to run sustainable forestry ventures is questionable when and where private enterprises show little interest in managing stands of trees so as to enhance future yields, establishing market contacts, and related pursuits. Even if the private sector has exhibited interest, preparing local populations to take on the latter tasks might not make economic sense.

One criticism of sector-specific human capital formation has been offered by Simpson and Sedjo (1996). In particular, they contend that a much larger area could be protected if the money used to subsidize specific activities like extraction of nontimber products were instead used simply to pay forest dwellers to keep natural habitats intact.

Just as there might be better ways to use scarce funds and technical assistance to save endangered ecosystems, sector-specific training may not be the most effective measure for raising forest dwellers' standards of living. By and large, those people derive more benefit from education that is more generally applicable, and which prepares them to take advantage of whatever well-paying employment opportunities that might come their way.

The spillovers generated by human capital formation, even of the sector-specific variety, cannot be ignored. Some of the skills that a worker acquires by training for and holding a job in, say, the ecotourism sector can be transferred to other occupations. Also, preparing people for work in nature-based tourism, forest management, and genetic prospecting can, under the right circumstances, diminish the cost of policing park boundaries.

The findings presented in chapters 4–7 indicate that sector-specific human capital formation is unlikely to lead to substantial income growth for large numbers of people living in or close to threatened habitats. Unless the skills forest dwellers acquire as a result of sector-specific training apply readily to other parts of the economy, the rationale for approaches to conservation and development that have become fashionable could turn out to be quite shaky.

II

The Economic Returns of Environmentally Sound Harvesting of Forest Products and of Nature-based Tourism

4

Harvesting of Nontimber Products

Collecting fruit, nuts, latex, and other nontimber products occupies at least some of the time of large numbers of people in the Amazon Basin and in other forested regions of Latin America. In northern Guatemala, for example, there are more than 7,000 chicle (*Manilkara zapota*) collectors, whose work generates $4 million in annual exports (Nations, 1989, cited in Salafsky, Dugelby, and Terborgh, 1992). At the time of the 1980 census, 68,000 households engaged in the collection of wild rubber (*Hevea brasiliensis*) in the Brazilian Amazon (FIBGE, 1982, cited in Allegretti, 1990). According to local newspaper reports, the harvesting, processing, and marketing of *açaí* and palm hearts currently employ nearly 30,000 people and generate an annual cash flow of up to $300 million in the Amazon estuary (Pollak, Mattos, and Uhl, 1995). In the whole river basin, 500,000 to 1,500,000 rural people derive a significant portion of their income from the extraction of Brazil nuts (*Bertholletia excelsa*), fruits like *aguaje (Mauritia flexuosa)*, and other commodities (Gradwohl and Greenberg, 1988; Schwartzman, 1989).

With very few exceptions, however, this work is not financially rewarding. All available research indicates that most extractor households are desperately poor, even by the modest standards of Latin America's forested hinterlands. Citing research carried out in Bolivia and in two places in Brazil, Browder (1992b) contends that households that collect nontimber products tend to be nomadic and illiterate, suffer from high rates of infant mortality, and are frequently in debt. In addition, they have often found it difficult to win governmental recognition of their natural resource rights when they are faced with the competing claims of loggers, ranchers, and agricultural colonists (Allegretti, 1990).

The Movement to Establish Extractive Reserves

It was out of struggles over land that the movement arose to create extractive reserves, defined as communal holdings where people support themselves by harvesting nontimber products. During the 1970s, cattle ranching expanded in the Brazilian Amazon, and large numbers of rubber tappers were dispossessed. In response, they began to organize themselves, with assistance from agrarian unions and Catholic social activists (Allegretti, 1990). By the late 1980s, the movement had a high political profile and had succeeded in attracting international backing.

Support for the rubber tappers was galvanized in December 1988, when ranchers killed Francisco ("Chico") Mendes Filho and two of his associates in the state of Acre, in western Brazil. Mendes had been an adept and charismatic union organizer whose international accolades included the Global 500 Prize of the United Nations. After his murder, the momentum to establish extractive reserves grew inexorable.

As far as collector households and their representatives are concerned, the new sort of landholding was never intended to exist for extraction of nontimber products alone. A Brazilian anthropologist who has worked with the rubber tappers since the 1970s has conceded that it is only to be expected that the inhabitants of extractive reserves will raise crops and livestock and cut down and sell timber whenever it suits them to do so (Allegretti, 1990).

Browder (1992a) confirms that crop and livestock production and logging are major sources of livelihood for the typical collector household. He also has warned those foreign environmentalists who have lent support to rubber tappers and other extractor populations against forgetting that the latter groups' agenda is primarily social—aimed at achieving legal recognition of their informal tenure in forested land—and only secondarily environmental.

The potential discord between international conservation interests and forest dwellers' desires for land rights was obscured, in a way, by a two-page article published in a prestigious scientific journal less than a year after the Mendes assassination. The article described a case study that appeared to demonstrate that environmentally sound extraction of nontimber products can be much more profitable than any economic alternative available to forest dwellers in the Amazon Basin.

The case study involved the estimation of sustainable production of *aguaje* fruit and other commodities on a one-hectare site near the town of Mishana, in the Peruvian Amazon. Output was multiplied by market prices in Iquitos, a nearby port city with more than 250,000 inhabitants located 30 kilometers downriver and to the northeast, and harvesting and transport costs were deducted. The resulting estimate of potential annual income was $422 per hectare. When the net returns generated by a selective timber cut every twenty years were taken into account, the present value of exploiting the site for fifty years was found to be $6,820 per hectare, assuming a real interest rate of 5 percent (Peters, Gentry, and Mendelsohn, 1989).

These estimates compare very favorably with the returns of other land uses, which usually create more environmental destruction. As Peters, Gentry, and Mendelsohn (1989) indicate, the potential income associated with nontimber-products extraction alone is two to three times greater than the per-hectare gross revenues that cattle ranchers in the Brazilian Amazon, for instance, are accustomed to earning. Likewise, the combined present value of incomes from all estimated extractive and logging operations turns out to be more than a dozen times the price at which rural real estate in the Amazon Basin normally changes hands. Of course, this implies that, in a perfectly free market, individuals engaged in nontimber-products extraction would end up owning land, and it would not be possible to cover the opportunity cost of other land uses.

The authors of the Mishana case study concluded that "without question, the sustainable exploitation of non-wood forest resources represents the most immediate and profitable method for integrating the use and conservation of Amazon forests" (Peters, Gentry, and Mendelsohn, 1989, p. 656).

Impediments to Economically and Environmentally Successful Extraction of Nontimber Products

Especially during the first year or two after Peters, Gentry, and Mendelsohn (1989) published their findings, many groups and organizations that were working to arrest tropical deforestation seemed to be convinced that the case for extractive reserves was irrefutable. But as Browder (1992a and

1992b), Redford (1992), and others were quick to stress, estimates of potential earnings at one site in eastern Peru could never be treated as conclusive proof that a way had been found to keep all rain forests intact while simultaneously raising forest dwellers' standards of living.

A number of issues were beyond the scope of the Mishana case study. For example, the production increases that would result if large forested tracts were dedicated to commercial harvesting of nontimber products would undoubtedly drive down prices, and hence extraction income. Peters, Gentry, and Mendelsohn (1989) left the analysis of this impact to other researchers. Likewise, they did not examine the marketing impediments that various entrepreneurs have encountered as they have tried to sell nontimber products in markets outside the Amazon Basin. Nor is it clear that postharvest losses were factored into the analysis. If, say, one-third of the potential harvest were never consumed or sold, then annual extraction income would drop from $422 per hectare to less than $200.

Something else to keep in mind is that Mishana is advantageously situated in two important respects. First, it is close to a sizable market; Iquitos is by far the largest urban center in the Peruvian Amazon. Moreover, the city is somewhat isolated, without good road connections to the outside world, and a large share of the population is descended from Indians and from the colonists who came to the area during the Rubber Boom (in the early 1900s); these people are therefore familiar with and like food and other things that can be gathered in rain forests.

Access to markets is critical to the commercial viability of nontimber-products extraction elsewhere. For example, much of Brazil's, and the world's, palm hearts are produced close to the mouth of the Amazon River (Anderson and Ioris, 1992). A crucial advantage of the area is that there are quite a few exporting firms in the nearby port of Belém. There is also a very strong market in that city for the fruit of the multistemmed *açaí* palm, which is the source of palm hearts in the Amazon estuary. Due to its perishability, *açaí* fruit is relatively expensive to transport. Hence, locational rents, of the sort depicted in Figure 3.1, are captured by producers situated near market hubs.

The second advantage that Mishana enjoys is that it is on a floodplain that happens to be dominated by trees bearing *aguaje* and a few other commercial fruits. Like market access, a location in a riparian area subject to periodic inundation is an advantage shared with the Amazon estuary, where *açaí* palms are abundant. Pointing out that only 2 percent of the Amazon Basin can be categorized as floodplain forests, Browder (1992a, p. 228) quotes Richard Howard, a former vice president of the New York Botanical Garden: "The most important point is how poorly the Peters et al. [Gentry and Mendelson, 1989] Peruvian Amazon hectare represents the tropical forests [of the entire region]; it is the rare exception rather than the rule." The Mishana investigators, it must be said, have been careful to avoid giving the impression that their site is representative of Amazonia as a whole (Peters, 1990).

Even if extraction of nontimber products at a site with favorable growing conditions has commercial potential, actually realizing this potential is far from a straightforward matter. Whenever demand is strong and harvesting practices are damaging, deterioration of the resource is all but assured in areas beyond the frontier of property rights consolidation. For example, one of the investigators in the Mishana case study is quoted as warning that wild fruit populations "are being rapidly depleted by destructive harvesting techniques as market pressure begins to build" (Vásquez and Gentry, 1989, p. 350).

Rapid expansion of extraction of nontimber products, in response to high demand for the plant in question, has often been followed by resource depletion and a collapse of the activity. *Cascarilla roja (Cinchona* spp.), which is the natural source of quinine, is a case in point. During the nineteenth century, the British decided to establish plantations of the species in India, so that there would be a ready source of medicine to combat malaria. But at least one of the botanists who was sent to gather seedlings along the western slopes of the Andes reported that his work was made difficult by the depredations of extractors, who often stripped medicinal bark in ways that killed trees (Spruce, 1970, pp. 240–241).

On the basis of detailed field observation of palm heart extraction, Pollak, Mattos, and Uhl describe why those who harvest nontimber products in what is in effect an open-access forest are not interested in conservation:

> Consider the extractor, who has to wade through mud, hacking away vines and other undergrowth, to reach the base of an *açaí* clump. Once there, he is inclined to cut all stems, even the small ones, before trudging through the forest in search of the next clump. . . . [There] are certain advantages to cutting smaller stems. . . . [In addition] delaying palm heart extraction

may mean losing the palm hearts altogether to another interested party. (1995: 376–377)

A sure sign that open-access predation is taking a toll on resource quality is that the average diameter of harvested palm hearts has declined dramatically since the early 1970s (Pollak, Mattos, and Uhl, 1995). It was during that period when much of the Brazilian palm heart industry relocated to Amazonia from southern Brazil, where destructive extraction had led to the depletion of *Euterpe edulis* Mart. (Ferreira and Paschoalino, 1987, cited in Pollak, Mattos, and Uhl, 1995). Unlike its Amazonian cousin, the latter plant is single-stemmed, and therefore does not recover easily after a harvest.

Aside from the depletion of open-access resources, there are other threats to the long-term viability of extraction of nontimber products. Agricultural domestication might occur. Also, synthetic substitutes can be developed. The most famous episode of domestication occurred during the early 1900s, when plants smuggled out of the Amazon Basin were used to establish rubber plantations in Asia. Since production costs were lower on those plantations than they were in Amazon rain forests, world prices fell and the South American Rubber Boom came to an end once Asian production began. Also, neither extraction of cascarilla roja in the wild nor plantation production proved to be competitive once synthetic quinine began to be manufactured.

If a close look is taken at successful examples of extraction of nontimber products, one is apt to detect many, if not all, of the key elements of agriculture, or agroforestry, to be more precise. For example, Pollak, Mattos, and Uhl (1995) have determined that the most profitable way to produce palm hearts in the Amazon estuary is in managed stands. In particular, management involves dispersal of *açaí* seeds and periodic thinning of undergrowth and canopy species. At current prices, applying such a system for twenty years yields a net present value of $119 per hectare, assuming a 6 percent real interest rate. Another benefit would be that processing facilities would no longer have to relocate every few years to progressively more remote locations, as a result of resource depletion (Pollak, Mattos, and Uhl, 1995).

Even though it creates little or no environmental damage, managing *açaí* stands, like what can be done with an aguaje "orchard," is not what some advocates of nontimber products extraction have in mind. However, those individuals should keep in mind that harvesting in unmanaged forests is rarely an alternative, if the purpose is commercial production as opposed to the gathering of forest products for household use. (The latter, of course, is an important feature of rural life in many forested areas of the world.) Reviewing historical experience in the Peruvian Amazon, Coomes (1995) observes a consistent pattern of harvesting far in excess of sustainable levels, followed by some combination of resource depletion, agricultural domestication (usually outside the region), and synthesis. Invariably, the earnings of extractor households have been marginal even during periods of peak activity. Once the peak has passed, those earnings dwindle, often evaporating entirely.

Extraction income continues to be very low. A financial analysis in one Brazilian community, for example, reveals that extraction's net returns at the household level are inferior net returns for intensive or extensive agroforestry (Anderson, 1989, cited in Browder, 1992a). In a survey of 164 rubber tapper households that he conducted in the Bolivian Amazon in 1981, Romanoff (1981, cited in Browder, 1992b) found that 96 percent were in debt and 71 percent suffered periodic food deficiencies. Significantly, Brazilian rubber tappers' earnings have been similarly meager in spite of the national government's efforts to maintain prices above international levels (Allegretti, 1990).

Vegetable Ivory Production in Western Ecuador

Insights into the true commercial potential of extraction of nontimber products. and how the income generated by this activity ends up being distributed, can be gained by examining the collection, transport, processing, and selling of vegetable ivory in western Ecuador.

Vegetable ivory, which is used mainly to make buttons, is obtained from the seeds of a hardy *tagua* palm species (*Phytelephas aequatorialis*) that grows in humid and seasonally dry tropical forests below an altitude of 1500 meters in northwestern South America (Acosta-Solis, 1944; Barfod, 1991). Ecuador began to ship *tagua* disks to button manufacturers in Italy and other countries around 1900. According to reports issued by the country's Central Bank, these exports peaked during the 1920s and 1930s, totaling nearly $20 million (in 1995 dollars) in 1925. Shortly after the Second World War, plastic buttons were introduced, which greatly reduced demand. For three decades beginning in the early 1950s, foreigners' purchases of vegetable ivory were negligible (Coles-Ritchie, 1996).

The *tagua* industry has rebounded in recent years. Exports rose from $1.5 million in 1987 to $3.5 million in 1988; shipments hit $6 million in 1991 and, for the time being, are staying at or above $4 million a year. (The true value of exports is probably much higher since under-invoicing of exports is a common practice in the country.) Italy continues to be the main importer, having purchased 81 percent of Ecuador's production in 1991. However, button manufacturers in other parts of the world are showing an interest in *tagua*, which is more appealing than plastic to many upscale clothing buyers. In addition, the small handicraft market for vegetable ivory has strengthened somewhat because of the ban on international trade in products derived from elephant tusks.

Modes of Production and Marketing Channels

Available research suggests that *tagua* can live for more than a century and that its seeds can lie dormant for well over a year (Acosta-Solis, 1944; Barfod, 1991). Practically all productive stands date back to the 1920s or 1930s and were established through secondary succession, not planting. Since the plant yields various useful products, including roofing materials and livestock fodder, few people were energetic about uprooting trees during the years when Ecuador exported almost no vegetable ivory. Maintenance, such as it is, involves little more than the occasional removal of dead fronds.

After being collected by rural households, *tagua* is sold to intermediaries. There are no significant barriers to entry in this sort of business and, as Southgate, Coles-Ritchie, and Salazar-Canelos (1996) have found, marketing margins are in line with what is observed in any competitive industry. By contrast, the top end of the domestic marketing chain is much more concentrated. Only a few firms slice dried *tagua* seeds into disks. The two largest processor-exporters accounted for roughly 45 percent of total shipments in 1991 and another three firms shipped 30 percent (Southgate, Coles-Ritchie, and Salazar-Canelos, 1996).

Lack of competition in processing and exporting has nothing to do with economies of scale in production since any business can expand capacity simply by installing more slicing machines, each of which is run by a single operator, and by enlarging the yard where raw *tagua* is dried. Instead, concentration is a consequence of barriers to entry on the marketing side. Historically, it has been all but impossible to get into the *tagua* exporting business without having good contacts among Italy's button manufacturers. A current initiative to promote sustainable *tagua* production in Ecuador (Calero-Hidalgo, 1992), which has been supported by Conservation International (CI), seeks to develop new markets. This is a challenge since clothing manufacturers that have not used *tagua* in the past need to be assured that large volumes of high-quality vegetable ivory will be available before they stop using buttons made of other materials.

A Survey of Vegetable Ivory Producers

In order to determine the distribution of net benefits from *tagua* production as well as possible responses to price changes, Southgate, Coles-Ritchie, and Salazar-Canelos (1996) conducted surveys of collector households, intermediaries, and processor-exporters in 1993.

The research began with the identification of firms engaged in the manufacture and international marketing of *tagua* disks. Interviews with representatives of five enterprises, with offices in Manta (a coastal port city), Quito (the national capital), or in both (see Figure 4.1), began in January and were concluded by May. In addition to yielding the information needed to estimate processor-exporter income, those interviews shed light on the domestic marketing of raw *tagua*, which was needed to plan surveys of intermediaries and collector households.

Instruments used in those surveys were developed in February and March of 1993. The Foundation for Training and Investment for Socio-Environmental Development (CIDESA), which is responsible for local implementation of the CI Tagua Initiative, and which has substantial field experience in the places where *tagua* is produced, was intimately involved in this work. It also arranged for a pretest of questionnaires in the middle of March, in which twenty collector households and three intermediaries in the province of Esmeraldas participated (see Figure 4.1).

After a minor modification of the questionnaires, surveying was carried out in Esmeraldas in April. Fifty-nine households were interviewed in three areas identified by CIDESA. Due to an interruption of funding, surveying of 22 households in the province of Manabí did not take place until November. Twenty-one intermediaries in both Esmeraldas and

Figure 4.1 Centers of *Taqua* harvesting and processing in Ecuador.

Manabí were questioned during the collector pretest and survey. The intermediary sample was comprised of truckers and other small businessmen operating in or close to the communities where members of the collector sample reside. As already pointed out, Southgate, Coles-Ritchie, and Salazar-Canelos (1996) found that intermediaries' earnings are similar to what one finds in other competitive markets.

Sample Descriptions The collector sample is representative of the rural poor who populate northwestern Ecuador and other areas where nontimber forest commodities are produced. The typical household comprises 4.4 individuals in Esmeraldas and 6.4 people in Manabí. Four-fifths of all the household heads were born in the community where they now live and most of the rest came from some other part of the same province. Education levels are low, adults in Esmeraldas and Manabí having completed 4.2 years and 3.0 years, respectively, of primary school.

No interviewee identified *tagua* collection as his main livelihood. Instead, 83 percent of the household heads described themselves as farmers. The other primary occupations reported were fishing and small-time commerce. Six percent of the sample worked at off-farm jobs, receiving approximately 5,000 sucres (equivalent in 1993 to $2.63) a day, plus lunch.

Most households in the sample have one *tagual*, averaging 9.7 hectares, where family members, and no one else, collect vegetable ivory. A third of the sample possesses two sites and 13 percent has three. Agroforestry is practiced on nine-tenths of the sites, with bananas, cocoa, coffee, and oranges (in Manabí) interplanted with *tagua*. There is no such interplanting on 14 percent of the Esmeraldas *taguale*s. Five percent of the Manabí sites were described by the households that exploit them as communal forests.

As discussed below, patterns of *tagua* collection in Esmeraldas are distinct from those in Manabí. In addition, there are minor differences in local marketing. Some households in Esmeraldas take their product to an intermediary in order to be paid a price that is, on average, 22 percent higher than what would be received at the farm gate. The typical intermediary is the owner of a small business, who contracts with truckers to deliver loads of vegetable ivory to processor-exporters. In Manabí, where the roads are better, slightly more than a third of the households sell directly to truckers; the balance is marketed to local merchants, in much the same way it is done farther to the north. There is no appreciable difference between farm-gate prices and what intermediaries pay for *tagua* delivered to their places of business (Southgate, Coles-Ritchie, and Salazar-Canelos, 1996).

Gains from Collecting, Marketing, and Processing Vegetable Ivory
Southgate, Coles-Ritchie, and Salazar-Canelos (1996) used household survey data to estimate an implicit daily payment for time spent collecting vegetable ivory.

The survey revealed that production is higher in Esmeraldas (400 pounds are obtained from a representative site during a single peak-season harvest) than it is in Manabí (200 pounds per harvest). Also, there are differences in labor inputs between the two provinces: 2.9 person-days per harvest in Esmeraldas versus 2.0 person-days in Manabí. Furthermore, producer-level *tagua* prices rose between April 1993, when Esmeraldas households were surveyed, and November, when interviews were carried out in Manabí. CIDESA employees and individuals involved in the vegetable ivory business suggest that prices went up in both provinces by roughly 80 percent—from 4,900 sucres ($2.58) to 8,800 sucres ($4.63) per hundred pounds in Manabí, for example.

Daily payments to labor employed in *tagua* collection were estimated for median peak-season harvests in Esmeraldas and Manabí. Calculations were made with both the lower prices that prevailed in April and the higher values observed in late 1993. In addition, all estimates reflect a small deduction for nonlabor inputs (e.g., the burlap bags used to carry *tagua* to market). As the estimates reported in Table 4.1 indicate, median daily returns to labor employed in vegetable ivory harvesting ($2.36 in Esmeraldas and $2.32 in Manabí) compared poorly with off-farm wages— $2.63 a day plus lunch (as noted above)—before prices rose. After the price increase, daily returns—$4.40 in Esmeraldas and $4.37 in Manabí—were above rural wages.

Whenever the daily returns for harvesting exceed the opportunity cost of labor, *tagua* resources have implicit scarcity value. Interestingly, the de facto owners of these resources seem willing to respond to positive *tagua* resource values by dedicating more effort to management. In particular, 70 percent of the collectors surveyed in Esmeraldas indicated that they would respond to higher prices by trying to increase the productivity of *tagua* stands. This could be done by pruning more carefully and frequently. (The other Esmeraldas households reported that additional earnings would be used to try to raise agricultural output [15 percent of the sample], to increase consumption [11 percent], and to pay for the education of children [4 percent].) One factor that has held down

Table 4.1   Daily Payments for Peak-Season
Tagua Collection in Ecuador

	Province	
Calculation Month	Esmeraldas	Manabí
At April 1993 prices	$2.36	$2.32
At November 1993 prices	4.40	4.37

SOURCE: Southgate, Coles-Ritchie, and Salazar-Canelos (1996): 76. Reprinted by permission of CAB International.

household-level earnings in the past has been limited competition among processor-exporters. These firms certainly appear to be highly profitable, as made clear in Table 4.2's account of what a representative enterprise would have earned, given April 1993's raw material prices.

The analysis of vegetable ivory processing and exporting carried out by Southgate, Coles-Ritchie, and Salazar-Canelos (1996) suggests that the industry has not achieved long-run equilibrium, with earnings well above the norm for the Ecuadorian economy as a whole. Understandably, then, entry into the industry has been taking place. Several new enterprises have begun to operate in Manta and Quito during the last two to three years. Without a doubt, greater competition helps to explain recent increases in raw material prices at the household level. Similar adjustments can be expected in the future as long as processor-exporters continue to earn supernormal profits. Interpretation of Table 4.2 suggests that, even if raw *tagua* prices rose to three times what they were in April 1993, profits would still equal 33 percent of revenues.

Since Southgate, Coles-Ritchie, and Salazar-Canelos (1996) completed their study, real *tagua* prices have not stayed at the levels that prevailed in late 1993. In November 1995, the prices producers in Manabí received were between 10,000 sucres ($3.45) and 12,000 sucres ($4.14) per hundred pounds. In Esmeraldas, where *tagua* seeds tend to be larger, producer-level prices have varied between 15,000 sucres

Table 4.2   Revenues, Costs, and Profits in *Tagua* Processing, April 1993.

Gross Revenues	$645,880
Sales of 225,000 pounds of disks	640,497
Sales of *tagua* flour and other by-products	5,383
Expenses	$255,079
Purchases of raw *tagua*	89,905
Wages and salaries	102,337
Administrative, electricity, and other costs	62,837
Profits	$390,801
Profits as a share of revenues	61%

SOURCE: Southgate, Coles-Ritchie, and Salazar-Canelos (1996): 77. Reprinted by permission of CAB International.

($5.17) and 16,000 ($5.52) per hundred pounds. The director of CIDESA attributes price weakness to diminished demand for premium clothing in Europe and other affluent parts of the world (Rodrigo Calero-Hidalgo, personal communication, 1995).

Extraction of Nontimber Products and Rain Forest Conservation

There are a few tropical forests in Latin America where harvesting nontimber commodities truly is commercially viable. In the floodplains of the Amazon River and its tributaries, for example, useful species, like *aguaje* and *açaí*, are not too widely dispersed, and access to major urban markets, such as Belém and Manaus, is relatively good (Goulding, Smith, and Mahar, 1996).

But in most places, the earnings associated with extraction of nontimber products are much more modest. Through the early 1990s, for example, the net returns for *tagua* collection were no better than the opportunity cost of unskilled rural labor in one of the poorest parts of South America. The same has been true of rubber tapping and other forms of nontimber-products extraction in the Amazon Basin, both now and in the past. By contrast, supernormal profits are probably being captured by the few Ecuadorian firms that process and export vegetable ivory. This is similar to what happened on a larger scale during the Rubber Boom. The Manaus Opera House, for example, is lasting evidence of the wealth that was lodged at the top of the domestic rubber marketing chain at the turn of the century.

It is fortunate that *tagua* palms suffer no damage as a result of harvesting, as has been the case for *aguaje* and *cascarilla roja* extraction. However, vegetable ivory collection is representative of commercial nontimber-products extraction in that it usually does not take place in locations that are biologically diverse. As previously mentioned, most *tagua* is harvested in stands that have emerged as a result of secondary succession. The users of these stands, needless to say, weed out other species that have no household or commercial value. The typical *tagual* bears little resemblance to an undisturbed primary forest. Instead, it represents a transition to agricultural domestication, which along with synthesis is what usually happens when demand for a forest product stays strong. Once domestication or synthesis is commercially successful, plants in the wild continue to have value only insofar as there is a need for repeated genetic improvement.

Outside of a few economic niches, characterized by good access to markets and relatively favorable harvesting conditions, the environmental wealth on which extractive activity is based is not economically scarce. This means that, even if resource rights were to be strengthened (which would help to forestall wasteful harvesting of some products), resource owners would probably not derive much rental income and their incentives to engage in intensive management (i.e., quasi farming) would remain weak. Likewise, the skills needed to gather nontimber products are not particularly unique. All this implies that few forest dwellers are likely to earn very much from nontimber-products extraction. That line of work, then, comprises a very shaky foundation for an integrated strategy for habitat conservation and local economic development.

Since peaking in the late 1980s, enthusiasm for the extraction of nontimber-products has ebbed considerably. Knowledgeable experts, like Browder (1992a and 1992b), have provided convincing evidence of the activity's limitations. As a result, it has become rare for the view to be expressed that harvesting products like *açaí* and *aguaje* can save tropical forests.

Likewise, the activity no longer plays as prominent a role as it once did in habitat conservation initiatives. Sponsoring development agencies have grown skeptical about the claims made by the few remaining enthusiasts. The typical venture aimed at promoting commercially viable harvesting of nontimber products is a small-scale one, aimed at taking advantage of a limited niche characterized by an existing and accessible market as well as favorable growing conditions. There is not room in Latin America for many such projects.

5

Environmentally Sound Timber Production

M any of those who hope to arrest the loss of threatened habitats in Latin America by promoting sustainable development of selected forest resources, like latex and *aguaje*, staunchly resist any form of timber harvesting. Their aversion to the latter activity is easy to understand since, for many years, most of the region's loggers have exhibited little if any concern for environmental impacts.

There is considerable scope, though, for containing the damage that results from logging. New systems that reconcile the growing and harvesting of commercial timber with biodiversity protection and watershed conservation are being developed. Better felling and skidding practices can be applied as well, thereby eliminating many of logging's long-term environmental impacts. In order to enhance regeneration after an initial selective harvest, extraction proceeds only after inventories and zoning, which includes the demarcation of sensitive areas that are to be left undisturbed. After that, vines are cut and directional felling is practiced in order to foster the regeneration of commercial species. Also, roads and skidder trails are designed and built with an eye toward avoiding soil erosion.

Research carried out by investigators affiliated with the Institute for Man and the Environment of the Amazon (IMAZON) reveals that the adoption of improved practices in the eastern Amazon is often hampered in forested hinterlands by low stumpage values, which are mainly a consequence of bounteous supplies of timber from lands that are not managed. That research is reviewed in the first part of this chapter. The second part contains an analysis of a project undertaken in the Peruvian Amazon, which yielded an innovative harvesting and management system. No attempt was made to continue applying the system after external support for the project ended. Outside technical assistance proved to have been vital. In addition, the attractions of sustainable timber production were diminished by the disruptions associated with guerrilla activity. Also having an impact were low stumpage values, which had to do with adverse governmental policies, inefficient marketing, as well as resource abundance.

The findings presented in this chapter suggest that, just as there are limited opportunities to keep natural habitats intact and to raise forest dwellers' standards of living by promoting the extraction of nontimber products, it is important not to exaggerate the financial appeal of environmentally sound logging. Like nontimber resources, timber is cheap in many places, which makes it difficult to justify the sorts of investment needed for sustainable management. Furthermore, projected increases in global roundwood prices will not be large enough to make a significant improvement in the commercial prospects of sustainable tropical forestry.

Logging in the Eastern Amazon

The challenge involved in sustainable development of standing tropical forests in Latin America is brought into sharp focus by a group of four studies carried out by IMAZON, a private, nonprofit research institute located in Belém. Each study addresses one contemporary mode of timber production or another in the state of Pará (see Figure 5.1), which was where 87 percent of all the Brazilian Amazon's roundwood was produced in 1988 (Verríssimo et al., 1992). Taken together, the four pieces of research furnish an idea of the forestry sector's evolution over time and across the landscape from an initial stage, in which timber is readily available and mechanization is minimal, to a situation characterized by higher capital intensity and incipient resource scarcity resulting from cumulative depletion.

Riparian Logging

One IMAZON study, carried out by Barros and Uhl (1995), addresses the oldest form of forest exploitation in the Brazilian Amazon. In the seventeenth century, the two investigators point out, high-quality timber began to be harvested

Figure 5.1 The Eastern Amazon.

close to the banks of navigable rivers, for shipment to Europe. The first sawmills and veneer factories were built in the Amazon estuary during the 1950s. These plants were internationally financed and powered by steam. Two species, *Virola surinamensis* and *Carapa guianensis*, accounted for most of the production, which was destined primarily for foreign markets. An active program of road building, which began in the 1960s (see chapters 2 and 3), has made it feasible to exploit forest resources in places away from navigable waterways. Nevertheless, timber harvesting close to the Amazon River and its navigable tributaries continues today.

One sort of riparian logging activity is not very different from what could have been observed during the colonial era. In *várzea* lands that have built up over time on floodplains, small independent teams, normally comprised of three men, fell timber and manually drag or float it to nearby riverbanks. There it is sold and assembled into log rafts that are pushed to mills by small boats. Clear evidence of low mechanization in this business has been obtained in a survey of sixty-three plants—both cottage facilities with circular saws and mills with band saws that employ thirty people, on average—that process *várzea* timber in the Amazon estuary and around Santarém (see Figure 5.1). In particular, Barros and Uhl (1995) found logs that had been cut with chain saws at just twelve of those sixty-three plants. The rest had been felled with axes.

Logging is more mechanized in riparian areas away from floodplains. The teams of five men or so that operate in these areas always use chain saws, and trucks are used to carry logs to rivers or roads. The scale of such operations is more than twice the scale of a typical *várzea* enterprise (2,311 cubic meters per annum versus 873 cubic meters) and annual production per person is higher in the former (492 cubic meters) than it is in the latter (265 cubic meters). The more mechanized system carries higher average costs as well. But the additional expenditures are compensated because timber harvested away from floodplains is typically of higher quality, and therefore fetches a higher price (see Table 5.1).

Loggers who operate in *várzeas* and adjacent dry lands tend to go about their business with little regard for forest regeneration. Barros and Uhl (1995) contend, though, that sustainable production is possible, especially in floodplains. As mentioned in the preceding chapter, these areas feature a less diverse array of species. In addition, *várzeas* are well stocked with commercial timber, growth rates are twice what is observed in dry land forests, and logging involves less damage to the canopy and ground because there are fewer

Table 5.1   Riparian Logging Costs in the Amazon Estuary, Early 1990s.

Cost Item	*Várzea* Enterprise (873 m³ per annum)	Dry Land Enterprise (2311 m³ per annum)
Labor expenses	$3.83/m³	$2.06/m³
Stumpage payments	2.90	5.85
Equipment and fuel for felling	0.00	0.48
Equipment and fuel for transport to river's edge	0.00	5.93
Combined costs	$6.73/m³	$14.32/m³
Payment received at river's edge	$9.00/m³	$18.00/m³

SOURCE: Data from Barros and Uhl (1995): 92.

vines connecting trees being cut down to those in which loggers have no interest (Barros and Uhl, 1995).

Barros and Uhl (1995) also identify three prerequisites for the sustainable development of *várzea* forest resources. First, the environmental knowledge that local people have must be complemented by training in inventorying, felling, extraction, and thinning techniques so that they can become effective resource managers. Second, their property rights in forested land must be stronger. Third, stumpage values must rise.

A small investment in technical assistance and modest policy and institutional reforms would satisfy the first two prerequisites for sustainable forestry development. However, meeting the third condition is apt to be more of a problem. A ban on log exports, instituted to protect wood processors, would have to be eliminated. However, it is possible, even likely, that the main impact of such a change would be the encouragement of logging in more remote areas. Barros and Uhl (1995) have found that the cost of transporting one cubic meter of harvested timber for a distance of 100 kilometers is as follows: $30.02, using a truck; $7.94, with a barge; and just $1.02 when logs are rafted. With transportation costs this low and with standing timber as abundant as the two IMAZON investigators report it to be, stumpage values in a free and competitive market are unlikely to rise very high, even in the most favorably located *várzea* forests. Instead, the main consequence of increased demand for stumpage, resulting from the lifting of the log export ban or from some other event, would be the acceleration of loggers' progress up the Amazon River and its tributaries, which Barros and Uhl (1995) report is already under way.

Extracting Mahogany in Primary Forests without Roads

The high cost of reaching the outside world and its markets has always hampered commerce in the Amazon Basin. To this day, economic activity remains largely confined to areas served by roads and navigable waterways, with only the most valuable forest products being taken from less accessible places. One such product is mahogany, a wood prized for its durability, workability, and attractive color.

Commercial harvesting of the species did not begin in earnest in the Brazilian Amazon until the 1960s, when the national road network started to extend into the region from the south. At that time, a standard pattern of operations for mahogany firms was established, which continues to hold as the industry has spread to the north and west. Instead of being comprised of large numbers of independent firms and households, each specializing in logging, timber transport, or processing, the industry consists of integrated businesses. A typical enterprise employs crews to search for and harvest mahogany; arranges to move logs to mills (which the same company owns); and sells the boards produced at its mills to foreign buyers, mainly in the United States and Great Britain (Verríssimo et al., 1995).

Since mahogany is so valuable, it is common for scouting and harvesting crews to operate at sites several hundred kilometers from their employer's mills, and far away from public roads and highways. This mode of operations is expensive, since it usually requires road construction and delivering supplies by air. In the second of the four IMAZON studies of the eastern Amazon's timber industry, Verríssimo et al. (1995) have examined the operations of a single integrated firm in southern Pará, to the west of the Belém-Brasília Highway and to the south of the Transamazon Highway (see Figure 5.1).

As indicated in Table 5.2, raw material costs incurred by the firm depend greatly on where mahogany is harvested in relation to processing facilities. When extraction takes place 600 kilometers away from the mill, stumpage payments drop to zero, because locational rents are negligible, and because there often is no owner around who can demand money for standing timber. By contrast, transport costs are more than an order of magnitude higher than what they are for a logging operation carried out at a distance of 50 kilometers.

For all intents and purposes, mahogany logging in primary forests amounts to mining. In inventories of three 100-hectare plots in southern Pará where felling and extraction had occurred in the recent past, Verríssimo et al. (1995) found virtually no representatives of the species with a diameter at breast height (dbh) of 10 to 30 centimeters; saplings were rare as well. Similar findings were obtained at four other sites in the region where mahogany had been extracted recently. The IMAZON investigators also cited evidence obtained in Mexico, Central America, and other parts of the Amazon Basin that showed poor regeneration of mahogany after harvesting, due to the absence of large seed trees and of extensive light-rich openings in the forest (Lamb, 1966; Snook, 1993). They conclude that more than 100 years might need to pass be-

Table 5.2   Mahogany Felling, Extraction, and Transport Costs in
Southern Pará, Early 1990s.

Cost Item	50 Kilometers from Mill	600 Kilometers from Mill
Stumpage payments	$70.00/m^3	$0.00/m^3
Reconnaissance and felling	4.60	4.60
Extracting logs from forests	23.80	$23.80
Transporting logs to mill	12.00	144.00
Taxes on logs	3.00	3.00
Interest charges	4.60	4.60
Combined costs	$118.00/m^3	$180.00/m^3

SOURCE: Data from Verríssimo et al. (1995): 48.

tween an initial mahogany harvest in natural forests and a
second cut.

Because of poor regeneration, the mahogany industry can
be expected to continue along its current track of venturing
ever farther into primary forests in search of undisturbed
stands of timber, at a progressively higher cost. As the second
column of figures in Table 5.2 indicates, the expense of delivering a cubic meter of timber to a mill approaches $200 as
the distance between logging and processing sites exceeds
600 kilometers. This expense is easy to justify, since mahogany boards fetch very high prices in world markets. But at
some point, the option of raising mahogany outside a natural
forest setting is certain to be commercially appealing. That
point appears to have been reached in southern Pará, where
a substantial amount of mahogany is starting to be raised,
often in fields where pepper or rapidly growing timber species are planted, too (Steven Stone, personal communication,
1996).

Mahogany does not grow very rapidly, which reduces
landowners' interest in raising the species in a plantation
setting. A more serious problem is the harm caused by the
shoot-borer moth (*Hypsipyla grandella*). In a financial analysis, Browder, Matricardi, and Abdala (1996) have found
that, if there were no insect damage, plantation production
would be profitable if stumpage could be sold for $150 per
cubic meter. Table 5.2's figures suggest that a mill should be
willing to pay such a price in order to avoid having to truck
in timber from remote forests.

Regardless of which path the mahogany industry takes in
the future—either continued ecosystem mining or tree farm-

ing—forest dwellers will be in a poor position to derive much benefit from it. Except for those fortunate enough to possess advantageously located stands, resource owners are unlikely to receive high onetime stumpage payments. Once harvestable stocks in the wild are sufficiently depleted and mahogany begins to be raised on plantations, the industry's interest in natural forests will relate only to its needs for fresh genetic material.

Logging along Active Frontiers of Colonization

When mahogany was being harvested primarily in areas beyond the reach of ranchers, farmers, and other settlers, environmental impacts consisted mainly of the eradication, for several decades, of commercial stocks of that same species, as far as Verríssimo et al. (1995) and other researchers have been able to determine. But otherwise, the IMAZON team has found that forest recovery, including the regeneration of some commercial timber species, was well under way a short time after mahogany loggers had left the scene.

The same sort of forest recovery can occur in more accessible places, as the third of the four IMAZON studies makes clear. Uhl et al. (1991) analyzed timber production in the late 1980s around Tailandia (see Figure 5.1), a town that was connected to Belém, which is about 200 kilometers to the north, with an asphalted state road in 1985. As in other places experiencing rapid settlement and expansion of the wood products industry, loggers working in the vicinity of Tailandia tended to be highly selective. At most, they were interested in twenty species; on average, just two trees, each with a useful volume of 6.2 cubic meters, were being extracted per hectare in the late 1980s. Even though felling and skidding resulted in substantial damage to trees left in the forest, including commercial species, timber was regenerating rapidly. On ten plots, each measuring 5 meters by 15 meters and located in separate harvest openings, an average of 14.3 small individuals (standard deviation, 6.7) from species that loggers in the region were harvesting were counted, although *Manilkara huberi*, which is the most commonly extracted kind of timber, proved to be rare (Uhl et al., 1991).

The depletion of commercial stocks is certainly a concern when and where the intention is to use natural forests as a long-term source of wood. Also, as indicated in chapter 1, logging can increase the risk of forest fires, sometimes dramatically (Uhl and Kauffman, 1990). However, there is another aspect of timber production in frontier regions that has more of an impact on the environment. As Uhl et al. (1991) documented in the case of Tailandia, 69 percent (in terms of linear distance) of all side roads suitable for vehicular traffic in the area were built entirely or partially by loggers. Constructing these roads, which were and continue to be a commercial lifeline for local farmers and ranchers, has often led to the conversion of forests into cropland and pasture.

That the upgrading of transportation infrastructure was linked to rapid agricultural land clearing around Tailandia had much to do with low timber values. In addition to whatever improvements loggers might have left behind, resource owners received only about $.80 per cubic meter for their stumpage. At prices this low, the present value of a few seasons of crop production or of a few years of cattle ranching exceeded the present value of a second timber harvest twenty years or so after loggers' initial intervention. For example, Uhl et al. (1991) estimated that crops harvested during a single season were worth $460 per hectare in the late 1980s. By contrast, the value of timber extracted from a secondary forest after twenty years of good regeneration would have been $770 per hectare, assuming that the price paid for felled timber in 1989, $15.10 per cubic meter, exhibited no real growth over time (Uhl et al., 1991). Obviously, managing forests after intervention so as to enhance future timber production would not have been remunerative, even at a very low real interest rate.

Uhl et al. (1991) contend that, during the late 1980s, there was some scope for raising timber prices, and thus for enhancing resource owners' interests in sustainable forest management in the vicinity of Tailandia. In particular, they argue that the mills that were purchasing timber and converting it into boards could have paid more than $18 per cubic meter for their inputs. Resource owners, they also observe, could have used the additional monies to pay for improved management of production forests. The IMAZON researchers point out that a typical mill's monthly operating and maintenance costs ($23,440), defined not to include depreciation and interest expenses, amounted to 81 percent of its monthly revenues ($28,800). Also the industry was not using raw materials very efficiently, two cubic meters being lost in processing for every cubic meter of finished product manufactured.

These observations do not add up to convincing proof that the wood-products industry's profits were excessive. Indeed, net returns were actually quite modest, particularly after allowances were made for depreciation and other capital charges. Profit margins were certainly exceeded by what has been earned by firms that process and export vegetable ivory in western Ecuador, for example. Also, the competitive market forces containing profits appeared to be strong, there having been forty-eight sawmills in and around Tailandia in 1989 (Uhl et al., 1991). In addition, low processing efficiencies were more a consequence, than a cause, of low stumpage values. After all, there was little reason to make the investments required to utilize a higher portion of raw materials when those materials were abundant, and therefore cheap.

Timber Production in More Settled Areas

As of the late 1980s, some of the Tailandia landowners surveyed by Uhl et al. (1991) were refraining from selling stumpage in the hope that prices would rise appreciably over time. These expectations could well have been ill-founded, if the findings obtained in the fourth IMAZON case study (Verríssimo et al., 1992) are any guide.

The setting for the latter research was the area around Paragominas (see Figure 5.1), which is on the Belém-Brasília Highway. This road having been in operation since the late 1960s, it is accurate to describe the region as an old frontier, one in which settlement processes, of the sort that were in full swing around Tailandia in the late 1980s, have largely run their course.

Much of what Verríssimo et al. (1992) have to report about timber production in the vicinity of Paragominas is similar to what Uhl et al. (1991) have found near Tailandia. Logging takes a toll on the forest; on average, twenty-seven trees greater than or equal to 10 centimeters dbh are damaged for every tree extracted. However, natural regeneration is robust, 4,300 seedlings and saplings of commercial species being registered per harvested hectare.

As one might expect in a place where both agricultural land clearing and depletion of timber resources have reached an advanced cumulative stage, timber values have started to rise. In 1989, when Verríssimo et al. (1992) conducted their field research, small logging firms, typically comprised of thirteen individuals with a pair of chain saws, a bulldozer and a log loader, as well as three trucks, were receiving $27.50 for a cubic meter of roundwood. This was 50 percent higher than what mills around Tailandia were paying for timber at exactly the same time (as noted above) and 50 percent higher than the price for logs extracted from dry lands close to navigable rivers a year or two later (see Table 5.1). Responding to higher prices, wood processing firms had chosen to operate at somewhat improved levels of efficiency. Instead of three cubic meters of raw materials being used to produce one cubic meter of output, as was the case in Tailandia. 47 percent of the roundwood going into a typical mill was emerging as finished product (Verríssimo et al., 1992).

Stumpage prices, however, were not much higher in the old frontier region than they were where resources were more abundant. As of 1989, a cubic meter of uncut timber in the Paragominas region was worth $1.84 (Verríssimo et al., 1992), compared to $.80 around Tailandia. One reason why prices remained low was that trucking in logs from other locations is not prohibitively expensive. In addition, processing facilities (e.g., small band-saw mills) can be moved from place to place without great difficulty. So can labor. In general, a high level of factor mobility diminished locational rents in the late 1980s, and has continued to do so to this day.

Since prices of standing timber have not risen very much in the vicinity of Paragominas, resource owners and loggers have been slow to adopt better management practices. Verríssimo et al. (1992) carried out an economic analysis of two such practices. The first is vine removal at least eighteen months before a harvest, which reduces collateral damage to commercial species left in the forest. The second is the practice of refinement thinnings, which promotes the growth of those same species after logging has taken place. Together, these two practices were estimated to cost $120 per hectare. But even this modest expenditure was found not to be remunerative under market conditions that prevailed in the late 1980s.

Since that time, deforestation has continued in the Paragominas region, which in turn has caused timber to become even more scarce. Stumpage prices currently exceed $5 per cubic meter (Paulo Barreto, personal communication, 1996), and mills are now paying more than $35 per cubic meter for roundwood. A recent IMAZON-sponsored survey of the region's wood-products industry indicates that several smaller mills, which can relocate fairly easily, have departed, presumably for areas where agricultural land clearing has not reached such an advanced stage. But less mobile plants, like veneer factories, are responding to higher prices by making the investments needed to raise processing efficiency. At the same time, at least some landowners are sufficiently sure of their property rights to begin considering forestry as a permanent land use option (Stone, 1996).

Technology for improved forest management is available, and several measures for softening logging's environmental impacts appear to be profitable. In another IMAZON study, Barreto et al. (in press) offer an economic assessment of various measures that are a part of planned timber harvesting. These include mapping out roads and skidder trails so as to reduce soil erosion, vine removal, and directional felling. Collecting data on a 105-hectare plot where all of this was done, and on a 75-hectare parcel where logging proceeded in the usual way with no planning whatsoever, the investigators have found that applying a plan to limit damages results in a 15 percent increase in logging crews' hourly output and reduces average machine time requirements by 27 percent to 37 percent. Furthermore, the failure to plan logging operations results in the waste of 26 percent of all felled timber—7 percent due to poor harvesting techniques, and 19 percent because logs that have been cut down are simply left in the forest. Between productivity impacts and the reduced waste of timber, planning results in financial returns of $3.70 for every cubic meter felled. These benefits do not reflect any payoff associated with leaving a better-stocked stand of trees behind after timber extraction has been completed.

Barreto et al. (in press) lament that no effective regulations are in place to oblige loggers to adopt improved practices. Under the Brazilian forestry law, harvesting and management plans are supposed to be prepared, officially approved, and applied in the field. But the public sector institutions responsible for enforcing this system are severely understaffed and poorly funded. In practice, any logging scheme featuring the removal of all large commercial timber, which really amounts to high-grading, receives governmental permission.

Like many others, the IMAZON investigators would like to see improved regulations and tighter enforcement. However, they also acknowledge that the adoption of sustainable forestry is seriously impeded because of a general lack of technical information and qualified personnel in the sector.

The Palcazú Forestry Project

IMAZON's program of research reveals that, except where deforestation has largely run its course and timber values have begun to rise, incentives for managing natural forests for wood production are weak in the Amazon Basin. Nevertheless, natural forest management has been promoted in many parts of the region. Under the auspices of one project, carried out during the 1980s in the Palcazú Valley of central Peru, an innovative harvesting and processing system was developed and applied, with the participation of local indigenous communities. That system was abandoned, however, soon after outside financial and technical assistance was withdrawn. A fundamental reason for the project's collapse was low stumpage values, which had much to do with resource abundance.

Setting and Background

Located approximately 400 kilometers northeast of Lima, the Palcazú Valley is typical of the places in Colombia, Ecuador, Peru, and Bolivia where the Andes give way to Amazonia. Elevation there varies dramatically, from 3800 meters above sea level, in the western reaches of the watershed, to 270 meters where the Palcazú River flows into a larger tributary of the Amazon River (Hartshorn, 1990). The climate is hot and wet. Average annual temperature is 23.6 degrees Celsius, and precipitation in a typical year, which features no pronounced dry season, is 6300 millimeters. Under these conditions, natural vegetation is luxuriant (Hartshorn, 1990).

Aside from a few places where fertile alluvial soils have built up over time, and which the indigenous population has used for generations as sites for raising beans, maize, and other crops (Benavides and Pariona, 1995), the Palcazú Valley is largely inhospitable toward agriculture. The red clay soils that predominate are extremely acidic, with a pH of 3.8 to 4.5; feature high concentrations of aluminum; and are prone to erode when exposed to driving rains. Moreover, they are highly leached, and consequently lack major nutrients like calcium, phosphorus, and potassium. Even less fertile are the white sandy-clay loam soils that border many rivers and streams. No more than 8 percent of the entire valley is suitable for crop production (Hartshorn, 1990).

Notwithstanding these limitations, agricultural colonization was the main thrust of central government policy for places like the Palcazú Valley for many years. Shortly after President Fernando Belaunde Terry, who had been deposed during a military coup in 1968, returned to office in 1980, his government announced plans to construct roads; to establish wood-processing and other industries; and to settle 150,000 colonists in the region, under the auspices of the Pichis-Palcazú Special Project (PEPP). Keen to support Peru's return to civilian government, the U.S. Agency for International Development (AID) promised funding and technical assistance.

The AID Project

From the outset, PEPP met with fierce opposition from the indigenous Yanesha (also known as Amuesha) communities and their Peruvian and foreign allies. Responding to these criticisms, and the cautionary advice of its own consultants (Smith, 1982, cited in Benavides and Pariona, 1995), AID decided not to back colonization. Instead, $22 million, including $4 million for technical assistance and project development, was allotted to the Central Selva Resource Management Project (CSRMP). A protected reserve was to be set up and managed and a system for sustainable timber exploitation was to be developed and applied. Environmentally sound agricultural production was to be promoted and public health services were to be upgraded as well.

As stressed by an expert affiliated with Costa Rica's Tropical Science Center (TSC), which provided technical assistance to the CSRMP's forestry component, the challenges of sustaining timber resource development in a place like the Palcazú Valley were considerable (Hartshorn, 1990). Government policies have accelerated agricultural land clearing and public sector institutions have had little or no capacity for furnishing useful advice to those with a potential interest in forest management. Low concentrations of commercial timber and high extraction costs were additional problems. In addition, there was a negligible level of understanding of tropical forest dynamics and the regeneration requirements of canopy tree species.

The CSRMP did not address policy issues or attempt a thorough overhaul of institutions engaged in forestry research and extension. Instead, primary emphasis was placed on developing and promoting an alternative to the usual pattern of unplanned high-grading of a limited number of species, which is what takes place throughout the Brazilian Amazon (as noted previously) and elsewhere in the American tropics. TSC investigators were convinced that an alternative approach was viable since, along with national and international demand for fine tropical woods, local and national markets exist for a wide variety of species. The implication of this is that the volume extracted from any given place could be increased appreciably, which would in turn diminish extraction costs and raise revenues (Hartshorn, 1990).

Due to the requirements of project implementation, there was no time for a thorough analysis of forest dynamics in the Palcazú Valley. However, the TSC assistance team was able to draw on observations of critical importance made in Costa Rica. Scientists at the La Selva Biological Station (see Figure 7.2) had documented that the growth of shade-intolerant species in gaps that open, when large trees are pushed over by the elements, is an intrinsic feature of tropical forest ecology. Half of all the tree varieties at La Selva, and 63 percent of the canopy species, require well-lit areas for regeneration (Hartshorn, 1978; Hartshorn, 1990). Gap-phase dynamics therefore became the central feature of the TSC management plan for the CSRMP (Hartshorn, 1990).

Experimentation carried out to identify the proper dimensions and orientation of gaps to be clear-cut began in 1985. A test strip 20 meters across proved to be too narrow for the emergence of large-crowned trees and it was determined that the growth of weedy species would be excessive on 50-meter-wide strips. It was decided to make production blocks, which vary in length from 200 to 500 meters, 30 to 40 meters in width.

Tree regeneration in these blocks, which are supposed to be logged every thirty to forty years, has been excellent for two reasons. First, neither burning nor farming, which destroy seeds, ever take place. Second, living seed sources are always nearby since strips that will be logged are always in the midst of intact forests. In addition, vines are removed before the harvest and enrichment thinning is done after tree emergence in order to enhance site quality. Also, soil loss is minimized by rotating logging sites and by running strips along topographic contours (Hartshorn, 1990; Hartshorn, Simeone, and Tosi, 1986; Tosi, 1986).

The TSC scheme called for extracting everything with a diameter of 5 centimeters or more. This represents a dramatic departure from standard practice in and around the Palcazú Valley. Like their counterparts in Pará, loggers in eastern Peru rarely cut down more than ten mature trees from a hectare of primary tropical forest. All other vegetation remains, often in a damaged state because of careless harvesting. Industry sources report that per-hectare extraction rates in the region are below 15 cubic meters and that high-quality hardwoods, cut with chain saws into crudely dimensioned planks, comprise most of the output (Southgate and Elgegren, 1995).

Of course, this pattern of forest exploitation makes sense where logging, transport, and processing costs are high. Electricity, for example, is considerably more expensive in eastern Peru than in other parts of the country. Since inefficient diesel-powered generators are the primary source of supply in the Amazon, prices there average $0.20 per kilowatt-hour, versus a national average of $0.05 per kilowatt-hour. Under these conditions, payments for electricity amount to as much as a fifth of wood processing costs and investments in processing capacity are minimal (Southgate and Elgegren, 1995).

In spite of adverse economic conditions, making use of timber with a small diameter was judged to be important to the functioning of the harvesting scheme based on gap-phase dynamics. Accordingly, a small mill was installed to convert timber of varying dimensions into different sorts of wood products: treated utility poles and fence posts, charcoal, and the sawed lumber normally exported from the region (Hartshorn, 1990; Simeone, 1990).

Another distinctive aspect of the CSRMP forestry component was the importance given to participation by indigenous communities. An early decision was made not to involve colonists, who had already converted most of their respective holdings to pasture and cropland and who lacked the social cohesion of the Yanesha. Work with the latter group began with participatory land use capability assessments. The Yanesha Forestry Cooperative, Limited (COFYAL) was established, and plans to extract timber from some forests and to set aside other areas as reserves were adopted democratically (Simeone, 1990). A major benefit that the native population associated with the project was stronger formal property rights; each of the fourteen indigenous communities in the valley received a communal land title (Benavides and Pariona, 1995).

Project Performance

Reflecting subsequently on the forestry activities he carried out with COFYAL, Simeone (1990) observed that outside technical assistance would be needed for many years if the production, harvesting, and milling scheme and marketing initiatives were to succeed. Poor performance of the system in the years immediately following AID's departure proved him right.

Foreign technicians and scientists paid by the U.S. government withdrew in 1989, largely in response to guerrilla activity in the vicinity of the Palcazú Valley, but not among the Yanesha themselves. Before that time, up to a dozen such individuals were active in the project. During the next four years, the World Wildlife Fund (WWF) and the Peruvian Foundation for the Conservation of Nature (FPCN), an environmental organization based in Lima, provided limited support. This allowed for four advisers, one each in administration, forest management, forest extension, and marketing, to continue working with COFYAL (Benavides and Pariona, 1995). But in 1993, even this assistance came to an end.

COFYAL's performance before 1993 compared poorly with what had been expected when the CSRMP was being designed, ten years earlier. Using data obtained from AID reports and other sources, Elgegren (1993) replicated the ex ante financial analysis of the forestry component. He estimated the base-case internal rate of return to be 20 percent, and also found that profitability was especially sensitive to changes in output prices and production costs. He also evaluated the cooperative's forestry operations in 1991, when harvesting took place on three strips with a combined area of 2.87 hectares. Significantly, he found that revenues ($5,491.83 per hectare) were lower than costs ($5,614.89 per hectare).

One reason why earnings were disappointing was that the prices received for COFYAL timber were low. On average, hardwood boards, which accounted for 40 percent of total production, were sold for $88.98 per cubic meter locally and for $135.59 per cubic meter in Lima. These prices were well below FOB border values, which exceeded $500 per cubic meter at the time (Southgate and Elgegren, 1995).

Uneven quality and marketing mistakes contributed to the low payments received for COFYAL output. On at least one occasion, for example, a buyer in the United Kingdom complained to the FPCN that there was too much empty space in shipping containers (Southgate and Elgegren, 1995). Needless to say, this drove down the payments received by the cooperative.

In addition, public policy tended to depress timber values. By the early 1990s, the Peruvian government was not regulating or taxing the export of unprocessed lumber. However, exporters were obliged to deposit foreign currency earnings with the Central Bank and then wait for several weeks to be paid back in Peruvian soles, at exchange rates set at the time of deposit. During 1991, when Peru suffered one of the highest rates of inflation in Latin America, this arrangement diminished the revenues received by wood exporters by 30 to 35 percent, on average (Southgate and Elgegren, 1995).

Depressed revenues were also a consequence of low production. Overall timber yields, which approached 45 cubic meters per harvested hectare, were three times what is extracted when normal logging practices are employed (see the earlier description of selective timber extraction). However, practically all of the difference was accounted for by the utility poles and fence posts (with yields of 55.40 and 188.85 units per harvested hectare, respectively, and with a combined volume of less than 30 cubic meters) manufactured from smaller timber; also, production of sawed tropical hardwood amounted to only 18.68 cubic meters per harvested hectare (Elgegren, 1993). The latter yield, in addition to being little more than what is obtained with standard extraction techniques, compares poorly with inventories of standing timber in the Palcazú Valley: 150 cubic meters of sawed logs per hectare and 90 cubic meters of posts and poles per hectare (Hartshorn, 1990).

It is revealing that the net economic losses ($123.06 per harvested hectare) incurred by COFYAL as it tried to implement the TSC system exceeded the net losses suffered by a private logging firm operating in adjoining lands. The latter, which were calculated taking all capital and operating and maintenance expenses into account and using data provided by the firm, amounted to $34.57 per harvested hectare (Elgegren, 1993).

Apparently cognizant of the financial advantages of the selective extraction techniques that prevail in central and eastern Peru and throughout the Amazon Basin, COFYAL decided to apply those same techniques on some of its lands at the same time that strips were being harvested in a way generally consistent with TSC guidelines. For example, only 46 percent of the timber that the cooperative produced for sawmilling in 1991 actually came from strips. The balance was extracted in much the same way that local loggers would have done it if given the opportunity (Southgate and Elgegren, 1995).

After WWF and FPCN support came to an end, the Yanesha disbanded the cooperative, which had never operated profitably. The latest word from the Palcazú Valley is that strip harvesting has ceased, as has the utilization of small-diameter timber. Both high-grading of timber and agricultural land clearing are taking place in the areas set aside for sustainable timber production by COFYAL (Benavides and Pariona, 1995).

Lessons Learned

Abandonment of the CSRMP does not mean that the efforts of AID, environmental organizations, technical assistance contractors, and COFYAL were entirely futile. The regeneration that is occurring on harvested strips suggests that the logging scheme developed under the project's auspices is biologically sound. Economic performance was indeed much less encouraging, although a conclusive test of feasibility was preempted by the disruptions caused by guerrillas and by adverse public policy.

To be sure, domestic prices for lumber and other tradable goods would have been higher, and incentives to apply the TSC harvesting and processing system would have been stronger, had exporters been able to choose when to convert foreign earnings into domestic currency. It can even be argued that, without price distortions, the opportunity cost of land dedicated to sustainable forestry would have been covered. Suppose, for example, that payments to COFYAL in 1991 had not been depressed by 30 percent (i.e., that revenues had been $7,700 instead of $5,500 per harvested hectare). With no improvements in the efficiency of timber extraction or milling, average annual income on a 40-hectare site, where a 40-year, TSC-style rotation was being followed, would have been $52.50 per hectare (equal to one-fortieth of the difference between $7,700 in revenues and $5,600 in costs). At a real interest rate of 10 percent, the present value of maintaining this income level indefinitely is $525 per hectare. If anything, this amount exceeds average farmland values in and around the Palcazú Valley (Elgegren, 1993).

But whether or not the TSC system is truly viable in the Peruvian Amazon remains in doubt. For the feasibility issue to be settled, one would have to examine not just the opportunity costs of land, labor, and other factors obtained locally, but also the scarcity values of management, marketing, and technical expertise brought in from the outside. COFYAL experience demonstrates just how critical the latter inputs were; operating the mill, for example, proved to be beyond the reach of the local community (Benavides and Pariona, 1995). The possibility that needs to be faced squarely is that, unless and until standing timber grows much more scarce, it will not make economic sense to devote a great deal of time or effort to forest management and related tasks in places like the Palcazú Valley.

Prospects for the Sustainable Development of Tropical Timber Resources

Covering the opportunity costs of *all* the inputs needed for forest management, timber processing, and the marketing of wood products is proving to be a major challenge in other sustainable forestry initiatives. Performance of a Bolivian project that is very similar to the CSRMP is a case in point— one of the participants argues that it would be impossible to practice good silviculture while at the same time carrying out processing and marketing operations without financial and technical assistance from the outside (Olivera, 1995).

Attracting management, technical, and marketing expertise to environmentally sound forestry ventures is sometimes made more difficult by public policy. The recent experience in Costa Rica of one of the TSC scientists who worked in the Palcazú Valley is instructive in this regard. That individual is currently trying to upgrade timber quality on a 600-hectare forested parcel, located in humid lowlands about 100 kilometers north of San José (see Figure 7.2). To promote regeneration at the site, where logging has taken place in the past, large specimens of the most valuable species will be left to stand for at least a few years so that they can serve as seed sources; only the smaller trees are to be harvested and extracted, preferably with oxen so that erosion and damage to remaining vegetation are minimized. However, the TSC scientist complains that this sort of innovation is being frustrated by public officials, who encourage high-grading, just as their counterparts in Brazil (Barreto et al., in press) and other Latin American countries do. In the specific case of Costa Rica, officials are accustomed to officially sanctioning selective logging (which normally involves the removal of all commercial timber larger than 60 centimeters dbh) and to granting permits to transport logs to mills in and around the capital city (which can process only raw material with a large diameter). He also complains that forest taxes, which amount to 10-to-15 percent of the value of harvested timber, diminish stumpage values, and therefore resource owners' interest in conservation (Joseph Tosi, personal communication, 1995).

The rules and procedures of international development banks and agencies likewise inhibit the free flow of skills and expertise to forestry ventures in Latin America and other parts of the developing world. Just about any project that AID proposes to carry out in a tropical forest setting must be preceded by a thorough environmental impact statement. Likewise, the Inter-American Development Bank (IDB) offers only limited support for the sector.

> In primary tropical forest settings, the Bank may support operations to enhance the ability of responsible agencies to manage forestry resources in a sustainable manner. However, the Bank does not finance commercial logging in these forests, nor the purchase of equipment for such purposes. (IDB, 1994: 34)

Foresters employed by the World Bank, which has similar requirements, indicate that they would be reluctant to spend a great deal of time working on a promising venture for developing timber resources in either a primary or a secondary forest. Their concern is that it might end up being too difficult to prove that a project is not running a substantial risk of unacceptable environmental damage.

One place where the ingredients for sustainable forestry seem to be present is Quintana Roo, in southeastern Mexico. Local communities there have organized to solidify their control over tree-covered land and to develop alternative marketing channels, so as to receive higher prices for timber (Bray et al., 1993). In addition, Quintana Roo holds major advantages over alternative sources of mahogany, which is the region's main wood product. For example, the cost of delivering logs to local sawmills has been estimated to be $80 per cubic meter (Richards, 1991), which is considerably less than the expenses that mahogany companies in the Brazilian Amazon face (see Table 5.2).

Even in Quintana Roo, though, there is no guarantee that environmentally sound development of timber resources can truly pay for itself. Germany's Agency for Technical Assistance (GTZ) has been providing advice and support since the early 1980s. Also, private foundations are supporting local forestry cooperatives' attempts to tap fledgling markets for tropical timber that one nonprofit group or another has certified as having been produced on the basis of sustainable forestry. But at the same time, Guatemalan production and exports, much of which are illegal, are exerting downward pressure on mahogany prices. Furthermore, a 25-year cutting cycle has been selected in Quintana Roo (Bray et al., 1993; Richards, 1991). Since all available research suggests that mahogany requires much more time to regenerate, the latter decision presumably reflects a finding that a longer cycle is unprofitable from the perspective of resource owners.

To admit that sustainable development of forest resources might not really be taking place in Quintana Roo would be discouraging for anyone pinning his or her hopes on a replacement for the wasteful and destructive logging that predominates throughout the American tropics. That AID's efforts in the Palcazú Valley were largely unsuccessful can be blamed on poor roads, guerrillas, and other exogenous factors. Also, the challenge to sustainable forestry is obvious in a place like Pará. Brazil's second-largest state is nearly as large as California, Oregon, Washington, Nevada, and Idaho combined and, since its timber resources are virtually boundless, market forces are stacked strongly against conservation. But the situation is different in Quintana Roo. The region is relatively accessible; social cohesion is better there than in many other places; and mahogany, which is widely sought around the world, grows well there. Thus, if outside technical assistance, which has been provided for more than fifteen years, will have to continue indefinitely in order to maintain a production system that does not allow for complete resource regeneration between harvests, the prospects for environmentally sound commercial forestry just about anywhere else must be considered dim, indeed.

Richard Rice, an economist employed by CI, which is involved in vegetable ivory production in Ecuador and is implementing a number of forestry projects in Latin America, has in fact concluded that promoting the sustainable development of timber in primary tropical forests is futile. Using data obtained in a CI project in the Bolivian Amazon, he has estimated that annual accumulation of commercial mahogany stocks in a well-managed stand is 4 percent. In addition, inflation-adjusted stumpage prices are increasing 1 percent per annum, which implies that the overall returns for mahogany management are 5 percent (McRae, 1997). Since financial instruments yield much more than 5 percent in Bolivia and many other countries, a resource owner can grow wealthier by harvesting timber and investing the proceeds than by practicing sustainable forestry.

Rather than trying to convince people to do something that is not in their best interests, Rice argues that it would be better to let forest owners proceed with selective extraction. Immediately afterward, land could be purchased at low prices, reflecting the depletion of standing commercial timber, and placed in protected reserves. Theodore Gullison, a tropical forest ecologist at London's Imperial College who has been working with Rice in Bolivia, suggests that such an approach might actually have less of an impact on biological resources than attempts at sustainable mahogany production do. As others have done, Gullison observes that large openings must be maintained in the forest canopy so that shade-intolerant mahogany saplings can mature. Furthermore, his own work in the field leads him to believe that biodiversity is being diminished due to the thinning of species that compete with mahogany in the same openings (McRae, 1997).

At the February 1997 meeting of the American Association for the Advancement of Science (AAAS), where Rice and Gullison presented their findings, the general proposition that alternatives to the sort of selective logging that one finds throughout the Western Hemisphere are neither rewarding for resource owners nor beneficial in terms of enhanced biodiversity was sharply contested. For example, Richard Donovan, who directs Smartwood—which is a program that certifies environmentally sound logging enterprises around the world—pointed out that mahogany extraction in Latin America's primary forests is much less intense than logging operations are in many parts of Asia and Africa, where it is common for half a dozen species or more to be exploited. He is sure that giving the latter operations free rein would be to invite disaster. Donovan also strongly doubts that extensive tracts of tree-covered land in the developing world can be placed under effective protection, even with the major infusions of cash that Rice appears to be contemplating (McRae, 1997).

Almost certainly, the prescription not to interfere with selective logging is not universally applicable. Nevertheless, it probably reconciles quite well with market conditions that will prevail for many years to come in Latin America's forested hinterlands. Drawing on an analysis of long-term trends in global roundwood markets, Sohngen et al. (1997) offer projections of stumpage values in various parts of the world. Of particular relevance to frontier regions in the tropics is their finding that timber demand growth, which is expected to be substantial, will be satisfied primarily through increased output in existing centers of production (e.g., the southeastern United States and the Pacific Northwest), and through a major expansion of tree plantations in the developing world, including Latin America. More timber will be extracted from the world's unmanaged forests, in boreal regions as well as the tropics. However, the latter output will continue to comprise a small portion of global production. Moreover, stumpage prices along the fringes of unmanaged primary forests in places like the Amazon Basin are unlikely to rise appreciably, since roundwood harvested in more accessible places, where management is more intense, will grow nearly as rapidly as consumption will (Sohngen et al., 1997).

These findings do not bode well for sustainable development of timber products other than fine tropical hardwoods in Latin America's primary forests. Market forces will not favor the sorts of management improvements that AID tried to introduce in the Palcazú Valley, and which CI has been promoting in the Bolivian Amazon. Sohngen et al. (1997) expect that unmanaged timber extraction will remain the norm. Without a doubt, local environmental damage, of the sort in Belize documented recently (Ito and Loftus, 1997), will often be severe.

The good news, though, is that waste and destruction will be confined mainly to the fringes of unmanaged forests. Other than those seeking mahogany and other highly valued species in trackless jungles, loggers will operate mainly in the vicinity of navigable rivers and passable roads. Where infrastructure remains undeveloped, natural vegetation ought to remain relatively untouched. The encouragement offered to the protagonist in the film *Field of Dreams* remains a clear warning about the consequences of highway construction in tropical forests: "If you build it, they will come."

6

Genetic Prospecting

Even though it is not particularly scarce, standing timber is the most valuable commercial asset found in tropical forests. For humankind as a whole, though, timber is probably of less importance than genetic resources. Although they occupy little more than 5 percent of the world's land surface, rain forests near the equator are the home of at least half of all plant and animal species (Myers, 1984; Wilson, 1988). Many of these species are threatened with extinction. As mentioned in chapter 1 of this book, Myers (1988) has identified several hot spots in Africa, Asia, and Latin America where tree-covered land rich in endemic flora and fauna is being encroached on rapidly.

None of the consequences of biodiversity loss is easy to quantify or to evaluate. It is conceivable that vital ecosystem functions, like hydrological cycle regulation, could break down if species diversity fell below some critical threshold. However, the risk of such an outcome is extremely remote in most places. By contrast, it is entirely reasonable to worry that avenues of research needed to cure cancer, AIDS, and other diseases might be hindered because too much biological raw material has perished irreversibly. It is worth remembering, for example, that the rosy periwinkle (*Catharanthus roseus*), which is the source of two uniquely effective drugs for treating leukemia, grows in the forests of Madagascar, which comprise one of the hot spots that Myers (1988) has identified.

Convincing though the story of the rosy periwinkle may be, circulating anecdotes is no substitute for thorough empirical study of the value of tropical forests as a source of inputs to pharmaceutical investigation. Without value estimates, it is hard to determine when and where deforestation is inefficient. Analysis of what the flow of biological raw material is worth to laboratories and other research venues is also needed to guide the articulation of legal and institutional arrangements for governing access to genetic information.

The best available evidence, which is surveyed in this chapter, suggests that there is no strong reason for pharmaceutical companies to pay a great deal to maintain supplies of genetic inputs from tropical forests. To be sure, society as a whole, either now or in later years, might attach a very high value to the lives saved because specimens can be gathered in the wild. But none of this has much of an impact in the marketplace.

Moreover, forest dwellers are not likely to derive substantial income from whatever medicinal products might be obtained from natural habitats. This is true even when they grow or collect a substance that, with limited processing, is consumed directly by people, rather than just being used in research. Consider the case of the *sangre de drago (Croton* spp.) tree, which grows in humid tropical areas throughout the Western Hemisphere, and which produces a sap that has been used for generations to cure various maladies. The product is starting to be sold in health food stores in Europe. Also, Shaman Pharmaceuticals Inc., of San Francisco, has been conducting clinical trials to test the safety and effectiveness of the substance as a topical treatment for drug-resistant herpes. Its usefulness in combating a respiratory virus that afflicts children is being investigated as well (Burton, 1994).

As a result of international interest, the *sangre de drago* market in Ecuador and several other countries has strengthened a great deal. A local entrepreneur who works with Shaman Pharmaceuticals reports that a hectare in the Ecuadorian Amazon with 100 trees (a 10-meter-by-10-meter spacing) yields approximately 300 liters of sap eleven years or so after the trees are planted, at a cost of $2.00 apiece, or emerge of their own accord (Douglas McMeekin, personal communication, 1994). As of January 1994, the producer-level price of the sap was $4.25 per liter. If that price were to hold steady in real terms over time, then the present value of a single planting and harvesting cycle would be a little less than $250 per hectare, assuming a real discount rate of 10 percent. That value is comparable to the opportunity cost of land in the region.

Harvesting a medicinal product like *sangre de drago* is about as remunerative as collecting *tagua, aguaje,* or any other nontimber product is in a South American forest. Since the net returns for the latter activity, which are modest, are examined in chapter 4, this chapter focuses exclusively on the value of tropical forests as a source of biological inputs to pharmaceutical research and development. The literature addressing that value is reviewed. Also described are the difficulties facing any party, be it an individual landowner or a national government, attempting to benefit by controlling access to habitats used in bioprospecting.

What the Pharmaceutical Industry Might Be Willing to Pay for Biological Raw Material

Various approaches have been used to evaluate the resources exploited by genetic prospectors. Laird (1993), for example, reports that payments for resource samples range from $50 to $200 per dry kilogram. But as Simpson, Sedjo, and Reid (1996) correctly point out, the people receiving these payments often have no property rights for the plants they collect, which means that prices would not fully reflect in situ values. To arrive at the latter, of course, labor and other expenses, which can be sizable, would have to be deducted.

A more commonly used approach for estimating the gross value of wild genetic resources is to draw on the pharmaceutical industry's experience with medicines derived from plants. In particular, the probability that an individual species will yield something commercially useful (i.e., the success rate) is multiplied by the returns associated with success. The resulting product suggests what an untested species is worth. But practitioners of this approach, and there have been quite a few, do not always distinguish between the gross and net values of a discovery. This exaggerates the returns to success since pharmaceutical research and development is usually very expensive. Overestimation also occurs because a distinction is not always made between average values of past discoveries and marginal values of new discoveries. This is a significant oversight because, as Aylward (1993) stresses, the gap between the two values is large and probably widening because the value of medicines derived from plants seems to be diminishing.

Farnsworth and Soejarto (1985) were among the first to base value estimates on industry experience. Farnsworth had analyzed the origins of drugs prescribed in the United States from 1959 through 1973, finding that plants were the source of one or more active agents in 25.4 percent of those drugs. Multiplying that share by the average price of a prescription in 1980 and by the number of prescriptions filled each year in the United States, Farnsworth and Soejarto (1985) concluded that the gross value of all prescriptions derived wholly or partly from plants was a little more than $8 billion:

> 25.4 percent: the proportion of all medicines with botanical ingredients
> × 4 billion U.S. prescriptions per annum
> × $8 per prescription
>
> ───────────────
>
> $8.128 billion: gross value of plant-derived prescriptions in the United States.

They also reported that, as of the early 1980s, only 5,000 species had been examined thoroughly and that, of that number, 40 were found to contain commercially useful medicinal ingredients. Applying an implicit rate of 1 research success per 125 species to one-fortieth of the gross value of plant-derived prescriptions, Farnsworth and Soejarto (1985) contended that the worth in the United States of an untested species is $1.63 million, in 1980 dollars:

> $1/125 \times 1/40 \times \$8.128$ billion =
> $1.63 million per untested species.

Principe (1989) carried out a study much like the one by Farnsworth and Soejarto (1985), but focused on all countries that belong to the Organization for Cooperation and Development (OECD), and not just the United States. Much of the data used by Principe was the same as the data Farnsworth and Soejarto had used, although he assumed much lower success rates, between 1 and 10 successes per 10,000 species. For a rate of 5 per 10,000, values in the OECD market, which is three times the size of the U.S. market, turned out to average a little more than $300,000 per species (Principe, 1989):

> 0.0005 success rate
> × $203 million per R&D success in the United States
> × 3 (ratio of the OECD market to the U.S. market)
>
> ───────────────
>
> $304,500 in the OECD per untested species.

This estimate, like the larger figure Farnsworth and Soejarto (1985) had arrived at a few years earlier, is a flawed indicator of in situ value because research and development costs were not deducted. In addition, it really amounts to an average historical value, not the marginal worth of an additional untested species.

In an evaluation of biodiversity values associated with a forest protection project in Cameroon, Ruitenbeek (1989, cited in Aylward, 1993) improved considerably on the notions of economic worth underlying the estimates provided by Farnsworth and Soejarto (1985) and by Principe (1989). The values of patented discoveries, rather than the prices charged for retail drugs, were examined. Aylward (1993) suggests that patent values may not be the best possible indicator to use in this sort of evaluation, but concedes that new discoveries are an appropriate measure of the output of genetic prospecting and of pharmaceutical research and development. To get at in situ values, specifically, it was assumed that only 10 percent of the value of new discoveries would accrue to the country in which the genetic material was originally found. Assuming that ten research successes, each worth $7,500, could be expected to result each year because of project implementation, the benefit to Cameroon would be $7,500 per annum (Ruitenbeek 1989, cited in Aylward, 1993):

$7,500 per research success
× 10 successes per annum
× 10 percent captured by the country where the sample was collected

$7,500 in annual biodiversity protection value captured domestically.

Aylward (1993) observes that, if the forest that is to be protected by the project contains 500 species, then the average annual return that the country collects because of species conservation is $15 per species:

$7,500 × 1/500= domestic benefits of $15 per protected species per annum.

The approach developed for evaluation of the Cameroon project has been applied, with modifications, in other settings. For example, Pearce and Puroshothaman (1995) sought to evaluate the biodiversity losses resulting from tropical deforestation. They used the range of success rates that Principe (1989) had identified—from 1 to 10 per 10,000—as well as rough estimates that he had also provided of the value of lives that would probably be lost in the United States because of the extinction of species with medicinal properties. Supposing as well that 60,000 plants are at risk in the world's tropical forests, which extend across 1 billion hectares, Pearce and Puroshothaman (1995) conclude that, from a U.S. perspective, the expected cost of lost genetic resources resulting from deforestation is between $0.012 and $21.00 per hectare:

60,000 tropical forest plant species at risk
× 0.0001 to 0.001—success rate
× $390 million to $7 billion per success
× 0.5 to 5 percent of the value captured by the host country
÷ 1 billion hectares of tropical forests

$0.012 to $21.00 in value of genetic resources per hectare.

A comparably wide range of species values has been estimated by Reid et al. (1993). They have modeled the species screening process in a more detailed fashion than other economists have done. In particular, they assume that there is a 1-in-10,000 chance that a biotic sample will contain a lead compound, and that only 25 percent of all such compounds will end up yielding commercial pharmaceuticals. Notwithstanding this insight, available data allow them only to conclude that the value of untested material is somewhere between $52.50 and $46,000 per species.

Aylward (1993) reports that more recent studies of biodiversity values reflect a better understanding of genetic prospecting and pharmaceutical research and development, and that the sort of conceptual errors that characterized earlier contributions to the literature are now being avoided. Not coincidentally, claims about what biodiversity is worth to drug companies and their customers have grown more modest over the years.

However, Simpson, Sedjo, and Reid (1996) argue that there is an upward bias in any estimate of the value of an untested species that is obtained simply by multiplying an expected success rate times the value of a new discovery, regardless of how accurately these two factors have been measured. Such an approach neglects the redundancy that can exist between one organism's genetic information and what another organism contains. Among the reasons why redundancy can occur is that the same useful compound might be found in more than one species. Also, different compounds can have the same curative properties.

Redundancy and the marginal value of untested species are closely interrelated. As Simpson, Sedjo, and Reid (1996) point out, a low level of redundancy implies that the chances are small that examining a given species will reveal anything useful. But marginal value is also low if the redundancy level is high since there is little chance of finding something new. Between these two extremes are levels of redundancy at which the probability of discovering something that is truly useful and unique is higher.

Taking redundancy into account, Simpson, Sedjo, and Reid (1996) offer a theoretical analysis of the linkages among screening costs, the value of a significant find, and the size of the sample of unscreened genetic material. Among other things, they find that, as the sample size increases, the marginal value of untested material grows small, even if screening costs are not very high. They also estimate an upper limit on the value of what they label the "marginal species." Using data from the pharmaceutical industry, they assume an average screening cost of $3,600 per sample, and that the average value of a research and development success is $450 million. Given some assumptions made about the distribution of useful compounds, the optimal success rate turns out to be 0.000012. At this rate, there is a 95 percent chance that a collection of 250,000 samples will yield a success, and the marginal value of the untested species is $9,431. However, this value is highly sensitive to the success rate. For example, if the rate falls by one-third, to 0.000008, then the marginal value dwindles to zero. The marginal value also declines if the success rate rises above its optimal level. At a rate of 0.00040, for instance, an untested species is worth only $67.

Simpson, Sedjo, and Reid (1996) show that the marginal value is also highly sensitive to sample size, the returns associated with a success, and other variables. In addition, they address the costs resulting from encroachment on natural ecosystems by combining estimates of the marginal value of untested species with a model of island biogeography (MacArthur and Wilson, 1967), which relates species extinction to habitat loss. They find that western Ecuador is the place where the pharmaceutical industry would be willing to pay the most to halt the conversion of forests into cropland and pasture. Myers (1988) considers the same region to be among the world's most active hot spots of threatened biodiversity since a huge number of species, many of them endemic, continue to survive there even though cumulative deforestation has reached an advanced stage. However, Simpson, Sedjo, and Reid's (1996) maximum estimate of the marginal value to the pharmaceutical industry of habitat protection in western Ecuador, $20.63 per hectare, is less than a tenth of what the region's farmers and ranchers routinely pay for cleared land (see chapter 2).

The remnants of natural habitats on three islands, where levels of species endemism are high, also feature relatively high biodiversity values at the margin: southwestern Sri Lanka ($16.84 per hectare), New Caledonia ($12.43 per hectare), and Madagascar ($6.86 per hectare). Next come three hot spots with biodiversity values just below $5.00 per marginal hectare: India's Western Ghats ($4.77), the Philippines ($4.66), and the Atlantic Coast of Brazil ($4.42). Elsewhere, saving one forested hectare is worth a couple of dollars or less to the pharmaceutical industry (Simpson, Sedjo, and Reid, 1996).

The industry certainly behaves as if tropical forests are not worth very much, in terms of the commercial (as opposed to the social) value of medicines they might eventually yield. Much has been made of the agreement that Merck and Company signed in September 1991 with Costa Rica's National Biodiversity Institute (INBio), under which the pharmaceutical manufacturer pledged to pay the latter agency $1 million per annum in exchange for plants and other raw materials, plus royalties on useful products derived from those inputs (Harvard Business School, 1992). However, one might speculate that Merck's interest in the deal had more to do with a desire to win favorable publicity than with sample collection, pure and simple. In any event, the amount of compensation involved is negligible in relation to what many commentators claim biodiversity values to be. (INBio has signed similar agreements, with comparable payments involved, with a few other companies.)

Other firms have tried to make similar arrangements in other countries. For example, in early 1995, Pfizer Inc. began negotiating access to Ecuador's biodiversity with the Ecuadorian Institute of Forestry, Natural Areas, and Wildlife (INEFAN). Originally, the company proposed to spend $998,000 on the acquisition and management of three small holdings and on a laboratory at which sample extracts would be prepared for export to foreign research and development facilities. INEFAN requested that the funds, instead of being used to buy land, be made available for the management of existing public parks and nature reserves. Also, both parties

agreed that a 1 percent royalty would be paid for any patented veterinary drugs derived from Ecuadorian materials
and a 2 percent royalty for human medicines (Roberto Ulloa
and Joseph Vogel, personal communication, 1996).

Pfizer stood to gain less public acclaim for saving species-
rich environments in 1995 than Merck had received four
years earlier. Pfizer even received a great deal of unfavorable
publicity, from the radical fringes of the environmental
movement in Ecuador and around the world. In response, the
firm decided in December 1995 to break off negotiations,
hinting that it would shift its operations to Brazil.

Controlling Access to Species-rich Habitats

As the actions being taken by INBio, INEFAN, and their counterparts in other countries demonstrate, national governments in the developing world are attempting to control
access to diverse genetic resources indigenous to the tropics.
INBio, for example, is trying to become a monopoly supplier
of the biological samples that Costa Rica ships to laboratories
around the world.

This interest in biological diversity contrasts sharply with
the total lack of regard for it exhibited by national governments not so very long ago. For example, the late Julian Steyermark, a U.S. citizen and a longtime Venezuelan resident,
put together and edited a comprehensive multivolume study
of Venezuela's plant life for the country's botanical institute,
which is a branch of the Directorate of Renewable Natural
Resources (DRNR). (When the work was published [Lasser,
1974], the institute's director, not Steyermark, was listed as
the series editor.) In 1980, the DRNR was transferred to a
newly created environmental ministry. Although this reor-
ganization was supposed to signal the national government's
enhanced environmental commitment, Steyermark's inven-
tory of genetic resources was soon mispalaced, and did not
emerge from a government warehouse for several years (Howard Clark, personal communication, 1996).

Although episodes like this cause one to doubt that developing country governments should be designated as the
sole custodians of the bounteous genetic resources that rain
forests and other tropical habitats contain, this is exactly
what is now being established. Local communities' ownership of traditional environmental knowledge is recognized
in Article 8J of the 1992 Global Convention on Biodiversity.
But the same treaty states that national governments have

sovereign rights over germplasm *and* its derivatives (e.g., proteins and alkaloids).

The legal and institutional structures that need to be erected for such a regime to operate effectively are monumental. Vogel (1994), who favors applying the existing system of intellectual property rights, including 15 percent royalties, to genetic material, acknowledges that a "gargantuan" data base would have to be developed and maintained for the expanded system to work. Such a data base would comprise far more than simple botanical and zoological inventories. In particular, detailed information on the spatial incidence of each species would have to be included, so that royalties generated by any plant-derived drug could be distributed fairly.

As Vogel (1994) emphasizes, nothing less than this sort of system would suffice for the emergence of robust markets for genetic information. Such markets, of course, would generate the sorts of price signals required for the efficient development of that information. Likewise, a well-articulated data base of the sort he describes would be a necessary, though not a sufficient, condition for the successful functioning of a cartel, made up of governments in tropical countries, that would control access to biodiversity in the wild—such a cartel, Asebey and Kempenaar (1995) indicate, is a possibility.

Investing in the detailed mapping of the world's biodiversity resources, so that a royalty system or a cartel can work, does not seem viable at present. The social benefits associated with using medicines derived from species collected in the wild might be substantial, reflecting, as they do, human beings' willingness to pay a great deal to extend life spans and to escape the ravages of disease. But until it can be conclusively demonstrated that those benefits exceed the legal and institutional costs of a royalty system or a cartel, neither of these two arrangements will be economically practical.

The Risk of Counting on Riches from Bioprospecting

Along with a film or two, a number of books and articles extol the medicinal wonders that come from plants and other biological resources found in tropical rain forests and known to people who live in this setting. Mark Plotkin, an ethnobotanist, has been particularly energetic about documenting what indigenous shamans in the American tropics know about preparing skin rash treatments, the curare extract used as a muscle relaxant, aphrodisiacs, and other products made from materials found in the jungle (Plotkin, 1994).

Plotkin contends that the pharmaceutical industry could come up with products of enormous value to all humankind by tapping forest dwellers' environmental knowledge; and the public imagination is captivated by stories of lone scientists working with tribal people to bring a cure, for AIDS or cancer, out of the rain forest, often doing this in spite of the depredations of loggers and miners. The tale of the rosy periwinkle, which nearly everyone who speaks or writes about biodiverse tropical forests feels obliged to repeat, is an essential part of this lore.

But the reality of pharmaceutical research and development appears to be quite at odds with romantic ideas about ethnobotany. Rather than sending scores of experts, either anthropologists who know about biology or biologists who are familiar with native cultures, into the field with video recorders, industry has chosen to invest heavily in the laboratory facilities needed for rapid assessment of large numbers of samples, many of which are microbial as opposed to being actual plants and most of which are collected without consulting local people. For example, Amgen Inc. boasts that it is now applying "newly emerging techniques in robotics and miniaturization" so that it can "synthesize and test thousands of potential drug compounds in a fraction of the time that would have been required in the recent past" (1997, page 17). These innovations comprise the cornerstone of a company strategy to have it evolve from a large biotechnology enterprise, with sales of two products accounting for more than 85 percent of total revenues, into a major player in the pharmaceutical industry.

Fundamental industry trends, as exemplified by the direction Amgen is taking, could be depriving humankind of cures for various diseases insofar as lines of ethnobotanical investigation that Plotkin (1994) has mapped out remain unexploited. But accepting industry trends as a given, one must concede that the models which economists like Aylward (1993) and Simpson, Sedjo, and Reid (1996) have developed and applied are reasonably accurate, as are their estimates of the marginal commercial values of untested species and species-rich habitats.

Those estimates suggest that, as a rule, genetic prospecting will not create a financial bonanza for countries with species-rich habitats. Assuming a 10 percent discount rate and a

1-in-10,000 success rate in pharmaceutical research and development, Aylward (1993) finds that the present value of royalty revenues provided by Costa Rica's 600,000 hectares of parks and reserves, which can be expected to yield 1,000 samples a year for testing, is approximately $4 million. This benefit estimate, which is consistent with what Simpson, Sedjo, and Reid (1996) have found concerning tropical hot spots' bioprospecting values, compares poorly with the opportunity costs of the same 600,000 hectares, which exceed $200 million (Aylward, 1993).

Certainly, the relationship between benefits and costs would be greatly altered if a "blockbuster" plant, like the rosy periwinkle, were discovered in a Costa Rican park. However, counting on such an event, which would have to be considered an exceptional case, is hardly a prudent financial basis for a habitat conservation strategy.

7

Nature-Based Tourism

Of all the economic alternatives contemplated for threatened habitats in the developing world, none appear to hold as much commercial promise as tourism. The business of accommodating people who wish to experience different planes firsthand has grown rapidly and, since the late 1980s, has accounted for 7 percent of all international trade in goods and services (Whelan, 1991). Formerly the esoteric pursuit of a few people, vacations taken for the purpose of viewing and learning about exotic flora and fauna in their natural settings now comprise an important part of the tourism market in many places (Boo, 1990).

Dramatic ecotourism growth in Costa Rica and the Galápagus Islands, which are the geographic focus of this chapter, has not been entirely free of adverse environmental consequences. Mountfort (1974) reports that careless photographers in the Galápagos occasionally interfere with the breeding of birds; and de Groot (1983) complains of people chasing marine iguanas (*Amblyrhynchus cristatus*), who must lie still under the tropical sun to recover body heat after emerging from the cold ocean. In addition, ships and boats that carry tourists around the islands routinely discharge garbage and sewage. Even at Costa Rica's Monteverde Cloud Forest Biological Preserve, which many reckon is one of the best-run protected areas in Latin America, soil sometimes erodes along visitors' trails. Also, tourists occasionally tap on the nests of resplendent quetzals (*Pharomachrus mocinno*), which are rare and finicky birds, to get them to take flight (Rovinski, 1991).

Except when threatened species, like the quetzal, are disturbed, impacts such as these are very localized, since the nature-based tourism sector does not make use of extensive tracts of land. For example, Huber (1996) reports that the total area managed by ecotourism enterprises close to Manaus, which is a mecca for foreign visitors to the Brazilian Amazon, is 4,000 hectares. This is a very small share of all the land in the region with ecotourism potential.

But if the environmental damages resulting from nature-based tourism are not of great concern, its local economic impacts are not very significant, either. It is almost always the case that, of all the money that visitors spend, only a small portion reaches local communities. Admonitions to increase local benefits, through planning, consultation, and other measures, are standard fare in the literature (Boo, 1990; Drake, 1991). But notwithstanding a few ballyhooed success stories, those measures have not been very effective in Latin America. Nor is it readily apparent that making the investments needed to enhance the local economic impacts of nature-based tourism would truly be efficient.

Local communities are not the only interest claiming more of the money spent by foreign visitors. In Costa Rica, the Galápagos, and elsewhere, maintenance and protection activities in the natural habitats that make ecotourism possible fall well short of what is required. In recent years, some of the money used to run parks and reserves, and to hire the guards needed to discourage encroachment, has been raised through debt-for-nature swaps and other mechanisms. Another way to put parks and reserves on sounder financial footing is to raise the fees paid by tourists and by the firms that serve them (Dixon and Sherman, 1990).

It is generally conceded that access to protected areas has been underpriced throughout Latin America, as it has been in many other parts of the world. However, caution must be exercised when fees are being readjusted for sites that have close substitutes. As David Simpson, an economist at Resources for the Future who has studied bioprospecting values in the developing world, puts it, "Every nice little hill and valley isn't going to be able to charge a lot for admission, since there are so many nice little hills and valleys that one could visit" (personal communication, 1996).

Simpson's warning is particularly relevant for those places in Latin America where price discrimination is being attempted. The potential gains from such a policy are illustrated in Figure 7.1, which shows the marginal costs (MC),

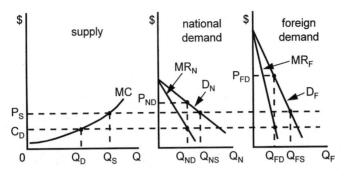

Figure 7.1 Price discrimination: The two-group case.

including environmental damages, associated with total patronage (Q) of a hypothetical park, and both national demand (D_N) and foreign demand (D_F) for access to that venue.

If everyone is charged the same entrance fee, P_S, then domestic and foreign visits will amount to Q_{NS} and Q_{FS}, respectively. Total costs will consist of the area under the MC curve between 0 and Q_S (i.e., the sum of Q_{NS} and Q_{FS}). These costs will be less than total revenues, defined either as P_S times Q_S or as the sum of P_S times Q_{NS} plus P_S times Q_{FS}.

But with price discrimination, a price is set in each market by finding the level of visits at which marginal revenues (MR) equal (a lower) MC. As shown in Figure 7.1, revenues collected in the domestic market are not much affected: P_S times Q_{NS} is about the same as P_{ND} times Q_{ND}. However, price discrimination results in more revenues being collected from foreigners, for whom demand is assumed to be less elastic; as can be seen in Figure 7.1, P_{FD} times Q_{FD} greatly exceeds P_S times Q_{FS}. At the same time, costs are lower if price discrimination is practiced since total visits are consequently diminished.

As can be readily appreciated from this conceptual discussion, the viability of price discrimination depends on there being significant differences among various groups' elasticities of demand for access to a park. There are some places where such differences exist, the Galápagos and the Monteverde Preserve in Costa Rica being prime examples. But elsewhere, national park services are attempting to apply the pricing strategy with little or no idea of what the true relationship between entrance fees and visits is. The results of the strategy can be disappointing.

The Ecotourism Boom in Costa Rica

Few places in the world are in a better position than Costa Rica (Figure 7.2) is to benefit from the interest that people have in visiting tropical habitats. The country is compact, with a land area of just 51,000 square kilometers. Within its borders, one can experience a wide variety of environments, ranging from beaches to jungles to mountains. Costa Rica is also home to a uniquely diverse mix of flora and fauna since it sits on the land bridge connecting North and South America. As of the early 1990s, 850 bird species and 208 mammal species had been identified in the country, as had countless insects and plants (Umaña and Brandon, 1992).

A serious effort has been made to promote travel to Costa Rica, which is easy to reach from North America or Europe. The response, as indicated by growing international arrivals, has been impressive, and tourism now makes a sizable contribution to the national economy.

Problems arise because all but a few of the communities near national parks and private reserves gain little from visits

Figure 7.2 Costa Rica.

made by hikers, bird-watchers, and others. Moreover, the continued success of ecotourism, which accounts for an appreciable share of international arrivals to Costa Rica, requires major improvements in transportation infrastructure and protected areas. The national government has made imaginative use of various financing mechanisms and has raised park entrance fees, especially those paid by foreigners. However, the issue of how to fund improvements is not anywhere close to being fully resolved.

Tourism's Recent Expansion and Economic Significance

Long-term trends in Costa Rican tourism cannot be plotted with a high degree of precision, mainly because of how people who arrived from neighboring countries during the 1970s and 1980s opted to describe the purpose of their visit to immigration officials. Many Central Americans, of course, were trying to escape civil strife and economic dislocation, but most claimed that tourism was their reason for spending time in Costa Rica.

There is no doubt, however, that arrivals from the United States and other wealthy nations, which had been growing slowly for several years, began to skyrocket in the late 1980s. As armed hostilities in Nicaragua and El Salvador drew to a close, North Americans and Europeans no longer sensed that a trip to Costa Rica involved any great danger. As a result, many more of them chose to vacation in the country. Indeed, 1989 was the first time when Canadian, U.S., and Mexican visitors first outnumbered the combined total of visitors arriving from Central America (see Table 7.1).

Costa Rican tourism continues to expand, albeit at a more modest pace. Since 1992, when arrivals from the United States were 23 percent higher than what they had been a year earlier, growth rates have fallen, from 14 percent in 1993 to 11 percent in 1994. Data collected by the Costa Rican Tourism Institute (ICT) indicate that the number of foreign air passengers landing in San José from January through September of 1995 was only 3.7 percent higher than arrivals for the first nine months of 1994.

Even though tourism might be growing more slowly, the sector continues to be a mainstay of the national economy. In 1994, the money spent in Costa Rica by international visitors was equivalent to 28 percent of the value of all the country's exports. Indeed, neither the value of the bananas sold

Table 7.1   Number of Foreign Tourists in Costa Rica, 1985 through 1994

Year	Canada, U.S., and Mexico	Central America	Europe	Total
1985	89,825	112,623	28,179	261,552
1986	93,105	106,825	29,026	260,840
1987	104,841	108,543	32,354	277,861
1988	123,551	124,728	41,396	329,386
1989	153,112	135,376	45,355	375,951
1990	191,284	139,913	57,177	435,037
1991	223,126	164,809	67,319	504,649
1992	274,061	187,790	88,301	610,591
1993	302,741	193,512	113,943	684,005
1994	332,602	221,384	129,580	761,448

SOURCE: Data from ICT (1995): 33.

overseas ($522 million) nor of coffee exports ($300 million) exceeded the $623 million in foreign exchange that tourism generated (ICT, 1995).

Importance of Nature-Based Tourism

Most international visitors choose to be in Costa Rica in December, January, February, or March, when precipitation is at an ebb. There are also a large number of arrivals in July and August, when North Americans and Europeans take a summer break (ICT, 1995). Especially during the dry months, when country roads are more passable, one can choose among a wide array of options for recreation.

Surveys reveal that, during the 1995 peak season, 82 percent of all visitors from the United States, which is Costa Rican tourism's primary market, were in the country on a tourist visa. Of this group, 76 percent went to the beach; 28 percent engaged in surfing, diving, and other water sports; 17 percent fished; and 11 percent went river-rafting. Many participated in one form of ecotourism or another: natural history (38 percent), bird-watching (35 percent), photographic tours (7 percent), and "tropical adventures" (38 percent). Some peak-season tourists from the United States also took advantage of their time in Costa Rica to learn Spanish, to attend a conference, to do some business, or to take in a museum (ICT, 1994).

The preceding data indicate only what foreigners end up doing while in Costa Rica, not their motivations for traveling there in the first place. The typical vacationer engages in more than one kind of activity, and it is important to gain other perspectives on the degree to which the recent tourism boom has been driven by visitors' desires to see rain forests, to watch birds, and so forth.

One alternative indicator of ecotourism's importance is the number of visits to protected areas, which rose dramatically in the late 1980s and early 1990s. As one official of the National Park Service (SPN) has documented, 404,342 foreigners entered Costa Rica's national parks in 1993, which was about five times the number recorded six years earlier (Bermúdez, 1992 and 1995, cited in Chase, 1995). There has been a decline during the past couple of years, in part because of entrance fee increases. However, patronage remains much higher than what it used to be. Rovinski (1991) and Aylward et al. (1996) report that similar changes have occurred over time in visits to privately owned areas, like the La Selva Biological Field Station and the Monteverde Preserve (see Figure 7.2).

Aylward et al. (1996) observe that the purpose of many of the visits to Costa Rican national parks might not be consistent with some of the narrower definitions of ecotourism that are used by some researchers and activists. The second-smallest park, Manuel Antonio, which only takes in 683 hectares, is consistently one of the most heavily used protected areas (it drew 94,102 international and 33,921 national visitors in 1994), mainly because it boasts spectacular beaches and because there are dozens of hotels and inns in the vicinity. The same year, 143,822 foreigners and 181,448 Costa Ricans went to Volcán Poas and Volcán Irazú. Although these two sites contain high-altitude cloud forests that are important for watershed protection and interesting to biologists, they attract tourists almost exclusively because each offers views of an active volcano.

Subtracting the patronage of Manuel Antonio, Volcán Poas, Volcán Irazú, and a few other sites, Aylward et al. (1996) estimate that there were 73,000 paid visits by foreign ecotourists to Costa Rican national parks in 1992, when 330,000 entries were recorded for the system as a whole. They contend that, when use of privately owned sites is taken into account, approximately 100,000 foreigners—one in every four vacationers from the United States, Canada, and Europe (ICT, 1995)—participated in nature-based tourism in Costa Rica in 1992.

Local Economic Impacts

Hard evidence concerning the local, as opposed to national, economic impacts of ecotourism is spotty. Enthusiastic claims are sometimes made about the benefits flowing to individual communities. But by and large, such claims are exaggerated.

It is widely accepted that the typical international vacationer spends a little more than $2,000 getting to, around, and back from Costa Rica. For example, a survey of 575 foreign visitors to the Monteverde Preserve, carried out in 1991 by the TSC, which owns and manages the site, revealed total expenditures of $2,207 per person, of which $1,273 were made somewhere within the country. For those individuals who indicated that seeing Monteverde was a prime motivation for being in Costa Rica, average overall and within-the-country expenditures were $1,961 and $1,131, respectively (Aylward et al., 1996).

Monteverde is indeed one place where ecotourism makes a significant contribution to the local economy. For the most part, relations are good between the preserve and the neighboring community, which Quakers from Alabama established in 1949. In addition, the town of Monteverde is pleasant, with clean restaurants and comfortable hotels, and the majority of foreign and national visitors to the area choose to eat their meals and to sleep there. Based on the finding, from the 1991 TSC survey, that the average duration of a foreigner's visit is three days, and using a recent ICT estimate of international tourists' average expenditures ($86 per day), Figueroa (1995) concludes that local spending by the 27,748 foreign visitors in 1994 was nearly $7,160,000. He also points out that more than half of what those visitors paid to enter the preserve and for souvenirs at its gift shop—$616,111 in 1994—goes directly to wages for local workers and to purchases of goods in the surrounding area.

But as a rule, money spent close to parks and reserves comprises a small portion of tourists' expenditures in Costa Rica. It is interesting to note that, notwithstanding the recent expansion of ecotourism, very little new lodging capacity has been constructed in the vicinity of ecotourism destinations. An ICT official reports that the spatial distribution of the country's hotel rooms is as follows: 31 percent in San José, 21 percent in other urban areas, and 31 percent along the Pacific or Caribbean coasts (Martín Quesada, personal communication, 1995). Even where there are lodging facilities of the sort that most foreign tourists (or at least those who are prepared to spend more than token sums of money) would care to patronize, those facilities are often self-contained, providing direct benefits to nearby communities only in the form of the low wages paid to waiters, maids, and other unskilled laborers.

For some protected areas, local economic impacts are practically nonexistent. For example, nearly everyone who visits Volcán Poas or Volcán Irazú does so on a half-day excursion from San José, which any hotel or guesthouse can easily arrange. Most visitors purchase little or nothing from people living around the park. Even at La Selva, the local returns are not all that great; Rovinski (1991) reports that the 13,000 individuals who visited the station in 1989 spent just $291,000 in nearby settlements.

Challenges to Continued Success

Now that international arrivals are no longer increasing as rapidly as they did a few years ago, impediments to the future expansion of Costa Rica's tourism sector, generally, and of ecotourism, specifically, are being given close scrutiny. If more tourists do not choose to visit the country and its parks, the chances are small that local economic benefits can grow appreciably.

Some barriers to expansion have to do with economic development in the broadest sense. Particularly outside the central valley where San José and most other cities are located, the quality of major thoroughfares and secondary roads is not what it should be. For example, the Panamerican Highway between the valley and Punta Arenas (see Figure 7.2), which is Costa Rica's main port on the Pacific Ocean, is a winding, bumpy, two-lane affair. Its only virtue is that it helps to steel the traveler for what awaits him or her on the byways that serve rural areas. These roads are not much better than country lanes in the poorest parts of Latin America. Just as they are a constant subject of lament among Costa Rica's farmers, rural roads that are muddy during the rainy season and rutted in dry months are roundly criticized by nearly every travel writer who visits the country (Ottaway, 1995).

Crime is another worry for foreign visitors. In 1994, nearly 4,000 international tourists were robbed in Costa Rica and a few suffered violent attacks of one sort or another. In January 1996, the kidnapping of two women—a German tourist and the Swiss owner of a travel agency in San José—near Puerto Viejo attracted global press coverage; along with demanding a $1 million ransom, the perpetrators threatened to attack a U.S. family and insisted on freedom for individuals imprisoned after the 1993 seizure of the Costa Rican supreme court building. Since the criminal targeting of foreign tourists could weaken the country's reputation for being Central America's lone haven of peace, policing is being improved in the capital city and elsewhere.

Habitat conservation is still another concern for the ecotourism industry. In past years, international development agencies and private organizations based in the United States and in other wealthy nations have provided large sums to Costa Rica for expansion and improved management of national parks and private reserves. Additional projects are in the pipeline. However, the country faces increased competition for funds from other parts of the world, including its Central American neighbors, where systems of protected natural areas are much less developed.

At the same time, the demands inherent in managing existing national parks appear to be straining the budgetary and human resources that have been allocated to the SPN. A benchmark for evaluating the adequacy of those resources is Monteverde. Over the years, the preserve has expanded to approximately 10,000 hectares. In 1994, the TSC employed 38 people there and spent more than 80 million colones ($509,327) on management (Figueroa, 1995). (This amount does not include purchases of merchandise for a souvenir shop and a refreshment stand.) By contrast, in 1995 the SPN assigned 58 employees and allotted 35 million colones ($198,999) to a region in the central highlands that takes in Volcán Irazú (2,309 hectares), Volcán Poas (5,600 hectares), two other national parks (with a combined area of 60,157 hectares), and a small national monument.

Support for the SPN and partner institutions in the public and private sectors will have to be increased substantially if proposals to expand protected areas in Costa Rica are adopted. For example, a team of consultants that worked for the World Bank has concluded that management of existing parks is successful in the sense that conspicuous fauna, with the exception of some large birds and a few other species, seem to be thriving. However, they advocate a 50 percent increase in the area under the direct control of the SPN, with special emphasis to be placed on the acquisition of lowland properties and on the protection of corridors among existing parks so as to allow for the unmolested migration of wildlife

(DHV Consultants BV, 1992). Even more ambitious would be the full implementation of the scheme to create nine conservation areas around the country, each to contain one or more national parks and to be locally administered (Umaña and Brandon, 1992).

That proposals of this sort are being given serious consideration suggests that the recent growth in ecotourism has made habitat conservation a much higher priority in Costa Rica than it would otherwise have been.

Pricing Issues

Over the years, many innovative approaches to financing habitat protection have been applied in Costa Rica. The country was one of the first to take advantage of debt-for-nature swaps. Through 1991, nearly $80 million of the country's foreign debt had been converted into $42 million in local currency bonds to benefit national parks and private reserves, at a cumulative cost of $12 million (Umaña and Brandon, 1992). Endowments to support the management of some protected areas have been established as well.

Since the very beginning of Costa Rica's ecotourism boom, economists have argued that habitat protection could be partly financed by raising park entrance fees. Several studies show that visitors' willingness to pay for access to protected habitats exceeds the nominal admission prices charged in September 1994. The fee increases that went into effect at that time seem to have been predicated on the assumption that demand for access to the country's national parks is price-inelastic across all major groups of visitors, especially foreigners who are now being asked to pay the most.

Baldares and Laarman (1990) made one of the first attempts to estimate the value that people place on visits to protected areas in Costa Rica. In a survey conducted by the SPN in 1989, when the entrance fee was 25 colones ($0.31), 860 visitors to Manuel Antonio, Volcán Poas, a national park on the Caribbean coast, and the Monteverde Preserve were asked what they thought daily payments should be. There was support for charging international visitors more than Costa Ricans were being charged, although foreigners preferred a smaller gap. Both groups favored raising the price for Costa Ricans to 50 colones and the international charge to 100 colones or so (Baldares and Laarman, 1990).

Two studies have addressed the value that visitors place on the Monteverde Preserve. In a travel-cost analysis, Tobias and Mendelsohn (1991) found that visitors' willingness to pay for access to the site amounted to about $12.5 million, which was more than an order of magnitude greater than what all of them paid for admission. Using data collected in the 1991 TSC survey mentioned above and contingent valuation techniques, Echeverría, Hanrahan, and Solórzano (1995) estimated that Costa Ricans were willing to pay $137 apiece, besides admission charges, to keep the reserve intact. On average, international visitors, who gain less from whatever watershed protection and local climatic stabilization results from conservation of Monteverde's forests, were willing to pay $119 apiece for the reserve's continued existence.

Findings such as these helped to justify the fee increases for foreigners that the SPN adopted in September 1994. The uniform daily price, 200 colones ($1.25), did not go up for Costa Ricans, but charges for international guests were raised markedly. The top fee, paid by those simply showing up at a park gate, was $15.00, paid in either dollars or colones, while someone making arrangements a day ahead of his or her visit was charged $10.00. Travel agencies were allowed to purchase tickets for $5.00 apiece. No discounts were made available to international students, which probably had a big impact on visits to privately owned sites, like the Monteverde Preserve.

Needless to say, a brisk trade in discounted tickets emerged quickly. Also, this evasion of the maximum fee was tolerated by some park administrators, particularly those responsible for areas that received relatively few visitors before and after the price increase (Chase, 1995). But the policy change still had a major effect on park use. During the previous peak tourism season, from December 1993 through March 1994, paid international visits had totaled 199,408; by contrast, only 113,461 foreigners were admitted to national parks during the next peak season, immediately after the price increase.

That a fourfold-to-twelvefold increase in daily admission charges for foreigners caused visits to decline only by 43 percent seems to indicate that international demand for access to Costa Rica's national parks is price-inelastic and, hence, the policy change raised revenues for the SPN. But Chase (1995) cautions that there might be a greater response to higher entrance fees in the long term. In particular, at least some tourists can be expected to choose destinations other than Costa Rica, and growth in SPN revenues might not be as spectacular as what some might have been expecting.

Aylward et al. (1996) contend that international tourists have not waited very long to revise travel plans. Since admission prices at the Monteverde Preserve—a token amount for Costa Rican students, $1.50 or so for citizens and residents of the country, $4.00 for foreign students, $8 for foreigners not on a package tour, and $16 for international participants in a tour—were not adjusted in the wake of the September 1994 increases in national park entrance fees, it should have experienced a major increase in visits. Indeed, from January through April of 1995, 9 percent more foreigners came to the area than those who came during the first four months of 1994. However, this was mainly because there was an 84 percent rise in visits by international students (who, to repeat, received no discount from normal admission prices at the national parks), which more than compensated for simultaneous reductions in the numbers of foreigners paying the $8 or $16 fee.

The impacts of park entrance fees on visits to Monteverde, and all that they might imply about substitute relationships between ecotourism in Costa Rica and ecotourism elsewhere, should not be exaggerated since prices for lodging and other services have been going up throughout Costa Rica as well. In addition, Chase's (1995) econometric analysis suggests that there is considerable scope for using price variations to direct tourists away from heavily used sites, like Manuel Antonio and Volcán Poas, to areas that could accommodate additional patronage.

However, the point that international tourists are cost-conscious should never be forgotten. The possibility that price increases might be causing more than a few international tourists to stay away from the country and its parks has not been ignored by the SPN, which modified its fee schedule in July 1995. Foreigners not buying tickets ahead of time will continue to pay $15.00 and the advance-purchase tickets to heavily visited sites will still cost $10.00; however, the price of advance-purchase tickets to other parks has been lowered to as little as $5.00.

Conservation, Tourism, and Local Interests in the Galápagos

As in Costa Rica, habitat protection and park pricing issues have become more important as tourism has expanded in the Galápagos (see Figure 7.3), the group of twenty-two islands and scores of smaller land formations, all of recent volcanic origin, 1,000 kilometers west of mainland Ecuador.

With a combined area of 800,000 hectares, the archipelago is home to a unique and largely endemic mix of flora and

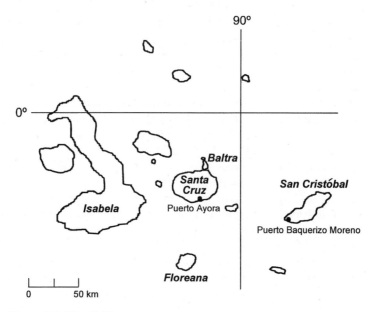

Figure 7.3 The Galápagos.

fauna. Among noteworthy creatures are the gigantic land tor-
toises (*Geochelone elephantopus*) for which the islands are
named, prehistoric-looking marine iguanas, various birds,
and sea lions (*Zalophus californianus*) and fur seals (*Arcto-
cephalus galapagoensis*), all of which show little or no fear
of humans. About 60 percent of the native plant and animal
species live nowhere else in the world (de Groot, 1983).

Of course, anyone who has taken a biology class knows
about the five weeks that Charles Darwin spent in the Galápagos in 1835. The observations he made there eventually
flowered into his theory that new species arise because of
natural selection (Darwin, 1859). The islands continue to be
an ideal natural setting for studying evolution. The thirteen
forms of Darwin's finches (*Geospiza* spp.), for example, provide textbook examples of rapid adaptation to variations in
the environment (Weiner, 1994).

Even by the time of Darwin's excursion, though, endemic
species were in jeopardy. It was the misfortune of adult tor-
toises, which have no natural predators, that they were able
to stay alive for up to a year in a ship's hold. Since there were
few other ways to have fresh meat during a long voyage, it
was a common practice for whaling ships and other vessels

to pass by the Galápagos to pick up a few large specimens. Intensive tortoise hunting took place for several decades during the nineteenth century, which caused some island populations to disappear.

For the most part, the threat to endemic plants and animals has involved less overt action by people. Mainland organisms, which easily supplant or prey on native flora and fauna, probably began to establish themselves in the archipelago not long after humans arrived on the scene, in the 1530s. The widely circulated notion that goats and pigs were introduced by sailors hoping to guarantee a food supply during subsequent visits is probably apocryphal. However, it is undeniable that domestic animals that have become feral and rats have adapted readily to local conditions. In addition, indigenous vegetation is losing ground to plants brought in, purposefully or inadvertently, from the mainland (Jackson, 1990).

The introduction of nonnative species, which continues to this day, is the lasting legacy of sporadic attempts by people to settle in the Galápagos (Southgate and Whitaker, 1994). Island ecosystems have been thrown far off their equilibrium as a result and, if the archipelago were simply left alone, many more indigenous organisms would become extinct. To avoid the outcome of a desolate landscape inhabited mainly by burros, goats, rats, cats, and dogs, rehabilitative action, which is expensive, is essential.

Conservation Initiatives

Ecuador first showed interest in protecting Galápagos wildlife as the centennial of Darwin's visit approached. Two of the country's earliest environmental laws relate directly to the archipelago. A nature sanctuary was established there in 1934 and the hunting of selected island species was prohibited two years later. However, these laws amounted to little more than an expression of virtuous intent since effective enforcement was precluded by the remoteness of the Galápagos.

International involvement was catalyzed two decades later by the United Nations Educational, Scientific, and Cultural Organization (UNESCO), which dispatched a field mission in 1957, and by the International Union for the Conservation of Nature (IUCN). The Charles Darwin Foundation was created in 1959 to take charge of protecting endangered species, and received official authorization to operate in the Galápagos. Its Charles Darwin Station, located on the outskirts of Puerto Ayora (see Figure 7.3), was dedicated in January 1964. The station coordinates scientific research in the islands, and its programs to revive land tortoise populations have enjoyed much success.

The presence of a research facility, combined with growing national and worldwide interest in habitat protection, led to the creation of the Galápagos National Park, the limits of which were demarcated in 1969 and 1970. Except for a little more than 3 percent of the land area previously occupied by the military, towns, and farms, the entire archipelago was placed inside the park. The first superintendent was appointed in 1972 (Southgate and Whitaker, 1994).

The Growth and Economic Significance of Ecotourism

In just about any accessible place in the world where natural scientists venture, tourists are sure to follow. Certainly, the Galápagos are not an exception in this regard.

An international scientific project mounted in 1964 demonstrated the feasibility of transporting relatively large groups of people to the islands. For the dedication of the Darwin Station, sixty-six scientists sailed from California and the Ecuadorian Air Force flew in dozens of government officials, diplomats, students, and special guests from the Ecuadorian mainland. After this event, a small but steady stream of visitors was attracted by the facility's presence.

In the late 1960s, international cruise operators sought to bring groups to the Galápagos, and contacted Ecuadorian travel firms to make local arrangements. By that time, the Ecuadorian Air Force had acquired larger planes, which made it possible to fly passengers to an airfield that the United States had constructed on Baltra during the Second World War. The travel industry soon linked up with the military's flight operations and the era of what would later be called ecotourism began (Southgate and Whitaker, 1994).

According to the Galápagos National Park Service (SPNG), the number of visitors has increased more than tenfold, from 5,000 in 1970 to more than 55,000 in 1995 (see Table 7.2). There has been no pronounced trend since the late 1980s in the number of Ecuadorian tourists. By contrast, visits by foreigners have grown steadily over time. Variation in the upward trend occurs when wars or civil unrest cause North

Americans or Europeans to fear traveling to Latin America (or elsewhere), and when additional capacity is authorized for the islands' fleet of cruise vessels. In the past, peaks in authorizations have coincided with changes in the national government, probably because incoming or outgoing officials feel obliged or able to respond positively to requests made by tourism operators during periods of transition.

Some of the economic impacts of Galápagos tourism are relatively easy to gauge. De Miras (1994) obtained a measure of total spending in 1993 by multiplying visitor numbers— 10,136 Ecuadorians and 36,682 foreigners (see Table 7.2)— by average estimated expenditures for each group: $505.61 and $1,336.82, respectively. The total value he obtained was $54,162,135.

This spending, however, represents just a part of the national economic significance of Galápagos tourism, since all foreign visitors must pass through Quito or Guayaquil on their way to and from the archipelago, and many of those people choose to spend a few days or weeks on the mainland.

Table 7.2   Number of Ecuadorian and Foreign Visitors to the Galápagos, 1979 to 1995

Year	Ecuadorian Nationals	Foreigners	Total
1979	2,226	9,539	11,765
1980	3,980	13,465	17,445
1981	4,036	12,229	16,265
1982	6,067	11,056	17,123
1983	7,254	10,402	17,656
1984	7,627	11,231	18,858
1985	6,279	11,561	17,840
1986	12,126	13,897	26,023
1987	17,769	14,826	32,595
1988	17,192	23,553	40,745
1989	15,133	26,766	41,899
1990	15,549	25,643	41,192
1991	14,815	25,931	40,746
1992	12,855	26,655	39,510
1993	10,136	36,682	46,818
1994	13,357	40,468	53,825
1995	15,483	40,303	55,782

SOURCE: Unpublished data from the Galápagos National Park Service (SPNG).

At least some of their expenditures during that time must be regarded as another direct benefit that Ecuador derives from the islands.

There is no denying that spending by visitors to the Galápagos is sizable. In a survey conducted in August 1995, the government's Ecuadorian Tourism Corporation (CETUR) found that 24 percent of all foreign vacationers arriving by air identified the islands as their primary destination; another large portion of the sample included a stop in the Galápagos in their itinerary. Moreover, foreign exchange earned because people from other countries visit or live in Ecuador has increased at a rapid pace in recent years, almost entirely because of the expansion of tourism. In the first six months of 1995, those earnings amounted to $144 million, which was equivalent to 6.6 percent of what the country received during the same period for its exports of petroleum, agricultural commodities, and other products (BCE, 1995, p. 53). Clearly, many millions of dollars, marks, and yen flow into Ecuador because of the Galápagos.

One reason why this is so is that the typical foreign visitor is affluent. Most of the people who make their way to the Galápagos are from rich countries; of the 457 tourists that Machlis, Costa, and Cárdenas-Salazar (1990) interviewed in the archipelago in July and August of 1990, 70 percent were from North America and Europe. These investigators found that virtually all the foreigners had at least some university education, and quite a few had graduate or professional degrees. One's country of origin and educational attainment are, of course, closely related to one's standard of living. Edwards (1991), for example, found that the average income for 360 foreign and Ecuadorian tourists interviewed in 1986 was $32,000.

The Galápagos are steadily becoming a premium nature tourism destination, visited mainly by people who are well off. Hotel and restaurant operators observe that fewer *mochileros*—young foreigners, who carry their belongings in backpacks (*mochilas*) and who watch their meager budgets with great care—are getting out to the islands. In addition, more Ecuadorian tourists visited each year during the late 1980s than during the early 1990s (see Table 7.2). Demand for visits among both these groups is fairly sensitive to airfares, expenditures on lodging and meals, park entrance fees, and other costs, many of which have been rising. By contrast, demand among affluent residents of wealthy nations appears to be price-inelastic, if the steady growth in their numbers in the face of rising costs is any indication.

Local Economic Impacts

Before ecotourism arrived on the scene, the Galápagos were one of the most isolated places on the face of the Earth. Since Ecuador laid claim to the islands, in 1832, settlement was attempted on various occasions. With some minor exceptions, though, the farmers, fishermen, soldiers, and romantics involved in these ventures all gave up after just a few years (Southgate and Whitaker, 1994).

The challenges settlers faced were considerable. Floreana and San Cristóbal (see Figure 7.3) are the only two islands with fairly dependable supplies of fresh water, and mainland markets for fish and other produce are distant. It is little wonder, then, that the Galápagus had only 1,346 inhabitants at the time of Ecuador's first national census, in 1950, and 2,391 a dozen years later, when the second census was conducted (INEC, 1992).

The infrastructure development that has accompanied ecotourism has made it much easier to travel to and from the archipelago. The airport on Baltra has been improved and another facility was opened in 1982 next to Puerto Baquerizo, the provincial capital. A third facility, with an 1800-meter landing strip, came on line in Isabela Island in February 1996.

To be sure, the possibility of finding employment in hotels, restaurants, or shops, or on boats or ships, also has stimulated immigration, which has been completely unrestricted since the Galápagos became a province in 1973. Annual population growth has averaged 5 percent, year in and year out, due mainly to arrivals from the mainland. The total number of inhabitants, concentrated almost entirely in Puerto Ayora and Puerto Baquerizo, was 9,785 when the last census was taken, in 1990 (INEC, 1992).

The attractions of the Galápagos remain strong for many potential migrants. Virtually every household in the two largest towns is connected to potable water and sewage systems and has electricity (INEC, 1992). Actual delivery of water and power may be subject to interruption, but no urban center along the Ecuadorian coast can boast of similar service. However, there are also disadvantages to living in the islands. Prices for food and other consumer goods, nearly all of which are brought in from the mainland, are high. Furthermore, many island residents have not found it easy to benefit from ecotourism.

Limited local earnings have to do primarily with how affluent visitors like to get around the archipelago. Different from *mochileros* and Ecuadorians, who often stay in hotels or inns and travel to different sites in small boats during the day, premium tourists generally prefer a cruise, either on ships that carry up to ninety passengers or on smaller craft that provide their six-to-twelve guests with more flexible itineraries. Operation of the larger vessels, in particular, requires few inputs from the local economy. The problem is compounded by a law that mandates a higher minimum wage and more restrictive work rules for Galápagos residents. Responding to that law, cruise ship operators do everything they can to hire crews on the mainland.

De Miras (1994) estimates that, of the $1,337 that an average foreign tourist spends getting to, around, and back from the archipelago, $102 goes into the local economy, in the form of hotel and restaurant charges, receipts for souvenir sales, and moorage fees. Additional benefits, such as wages and salaries paid to park guards and cruise ship employees, are not very large. A sophisticated analysis, of the sort needed to estimate the total impacts on the local economy resulting from these stimuli, has not been carried out in the Galápagos. One would be surprised, though, if those impacts amounted to as much as one-third of what well-off foreigners are spending.

The Role of Public Policy

Local impacts could be enhanced by changing government policy. In particular, new development initiatives in the islands could be funded by raising the taxes and fees paid by tourism operators and the people they serve. There is some scope for doing exactly this since a few operators appear to be earning supernormal profits. For example, the two airlines that operate in the archipelago—one private and the other linked to the armed forces—have exploited their monopolistic position to engage in exactly the sort of price discrimination depicted in Figure 7.1. In December 1995, for example, when the price charged Ecuadorians for a round-trip ticket was 558,000 sucres ($192), international passengers were paying $377 for exactly the same service.

The government could also collect more money directly from tourists. Edwards (1991) has carried out research which shows that sizable revenues are being foregone because of current pricing arrangements, in which every visitor pays the same price regardless of the length of his or her stay. He estimated that imposing a daily fee of $214 in 1986, when the flat price for foreigners was $40, would have caused average trip duration to decline by 50 percent. Regulatory controls on the number of visitors could have been dispensed with entirely without any net environmental impact because the doubling in visitor numbers that would have occurred would have coincided with the reductions in trip length associated with charging daily fees. In addition, park revenues would have risen from $40 to $770 per person (Edwards, 1991).

The government has never seriously considered instituting a daily price. However, entrance charges were raised substantially in 1993, from $40 to $80 for foreigners and from 600 sucres ($0.35) to 12,000 sucres ($7.00) for Ecuadorians. In addition, there was an increase in the fees levied on passenger vessels. As a result of many years of high inflation, the annual charge (*patente*) assessed on each passenger berth had fallen below $10. Large ships, capable of carrying ninety guests, only had to pay about $600 apiece in 1992. By contrast, surveys carried out by Bruce Epler, who was a researcher working in Ecuador for the University of Rhode Island at the time, revealed that vessels of that sort were earning as much as $4 million in annual gross revenues (personal communication, 1992). No data are available, but no knowledgeable observer thinks that annual operation and maintenance expenses exceed $2.5 million for any ship carrying passengers around the Galápagos.

INEFAN, which holds the administrative responsibility for all the country's parks, responded to these findings by raising berth fees. In general, vessels equipped to accommodate foreigners in reasonable comfort pay the "luxury" rate of $200 per berth per year. Annual *patentes* of $100 or $150 are charged tourist craft that either are smaller or lack an escape boat or advanced fire control measures. Day boats pay $30 per annum.

As a result of the higher entrance fees and *patente* increases, revenues have risen substantially. In 1993, $2.2 million were collected. Because of growth in international arrivals, 1994 and 1995 revenues were $3.2 and $3.7 million, respectively. The portion of this money going to the islands, however, has been small. INEFAN and predecessor agencies have always used Galápagos tourism to help pay for the management of mainland parks, none of which comes close to being financially self-supporting. In 1993, for example, the SPNG kept approximately 30 percent of the $2.1 million it collected; the rest was spent in other parks.

To capture a larger share of SPNG revenues, local governments will have to develop specific and viable proposals for development projects—which has not happened so far. It could also be that they will no longer have to compete just with INEFAN for funds collected in the Galápagos. In March 1996, the Ministry of Finance announced that, after consideration of INEFAN's requests to budget 3.5 billion sucres (worth approximately $1.2 million at the time) for the SPNG and 5 billion sucres ($1.7 million or so) for mainland parks, only 850 million sucres (roughly $300,000) and 2.7 billion sucres (about $900,000) were actually to be allocated during the year. This action could have resulted from the need to settle bills incurred during the small war that Ecuador fought with Peru in early 1995, or to pay for other projects to which the government had assigned a higher priority. However, the central government's decision to capture a higher portion of the entrance fees and *patentes* collected in the Galápagos might signify that, in the future, local governments, the SPNG, and mainland parks will face new, and more powerful, competition for scarce financial resources.

Finally, it must be conceded that the fee increases instituted by INEFAN since 1992 might have been detrimental to islanders' immediate financial interests since higher fees, along with increases in airfares and other prices, have helped to discourage visits by *mochileros*, Ecuadorians, and other less affluent tourists who tend to patronize on-shore facilities. Higher costs, it is fair to say, have accelerated the gentrification of Galápagos tourism, which is diminishing the earnings of many of the islands' hotels, restaurants, and shops.

Crises Coming to a Head

Benefiting in only a limited way from international visitors' interest in the Galápagos, residents have been quick to seize whatever economic alternatives present themselves. At the same time, they have responded eagerly to calls for political action to close the gap between what they think they should receive from ecotourism and what actually comes their way.

Some alternative economic activities, like fishing, involve substantial environmental damage. Since the government has virtually no capacity for effective regulation, intense harvesting has led to severe depletion of various species. The latest episode of boom-to-bust exploitation of an open access fishery began in 1991, when people and firms responsible for depleting sea cucumber (*Isostichopus fuscus*) populations along the coast of mainland Ecuador started to transfer their operations to the Galápagos. The sea-bottom-dwelling relatives of the starfish, which are sold as an aphrodisiac in China and elsewhere in Asia, proved to be particularly abundant in the waters around Isabela. Furthermore, the island is far beyond the reach of the SPNG, the Subsecretariat of Fishery Resources (SRP), or the Ecuadorian Navy. As a result, individual fishermen and the oceangoing transport ships that routinely pass by to make purchases have gone about their business with little official interference. During the early stages of the sea cucumber boom, a fisherman on Isabela could make as much as $700 a week, which stimulated a major increase in harvest rates as well as migration. By the middle of 1992, daily production for the entire archipelago was averaging 70,000 to 110,000 sea cucumbers. In 1994, when a ban on harvesting was declared (but never really enforced), approximately 116 fisherman had settled on Isabela (Zador, 1994).

The collapse of natural stocks due to overfishing could be a catastrophe for the entire island ecosystem. There is no way to know, for example, which kinds of birds might depend directly or indirectly for their survival on sea cucumbers, which are the most plentiful form of marine life in the archipelago. Fishermen, of course, show little or no regard for these consequences and have vigorously resisted any attempt to regulate their activities.

Islanders (both those born in the Galápagos and those who have lived there for several years) and sea cucumber fishermen are largely, though not entirely, distinct populations. In addition, many islanders either have a stake in the success of ecotourism or hope to have one in the future, so they have some interest in ecosystem integrity. However, they have given enthusiastic support to Eduardo Veliz, who was elected in 1994 to represent the archipelago in the National Congress, and other political leaders who are demanding a larger share of tourism's benefits for the islands. For example, a general strike was called in September 1995 to protest the Ecuadorian president's veto of Veliz's bill, which among other things would have increased automatic appropriations by the central government to local governments and projects and forced all tourists in the Galápagos to spend at least one night ashore. The strike succeeded in closing most private businesses and interfering with access to the Darwin Station (Lemonick, 1995).

Responding to the concerns of Galápagos residents, as the Ecuadorian government has committed itself to doing, involves some serious conundrums. For one thing, undertaking more projects to benefit local communities could stimulate additional migration to the archipelago. That such migration would dissipate the benefits accruing to individual households is widely recognized. Even Veliz has expressed some sympathy for coupling migration controls, which his predecessor supported (Brooke, 1993), with increased financial support from the central government. Such support, of course, could come at the expense of budgets for habitat management and species protection.

Adding to the difficulty of choices facing the Ecuadorian government is the international scrutiny of what takes place in the Galápagos, which has intensified in recent years. Visitor numbers fell sharply in 1995 in response to the islanders' strike. Since then, ecotourism has rebounded, but now UNESCO appears poised to declare the archipelago a "world heritage in danger." Responding to the damage that such a declaration would do to national prestige and foreign exchange earnings, Ecuador's interim president recently issued a decree addressing various issues, including pollution, migration, illegal fishing, and the introduction of species. Though encouraging, this action cannot take the place of comprehensive legislation addressing conservation and local development issues. After years of debating various proposals, the Ecuadorian Congress has put the future of the Galápagos back on its agenda. However, there are no guarantees regarding either the content or timing of a new law (*Economist*, 1997b).

Ecotourism, Habitat Protection, and Local Economic Development

There is no denying that Costa Rica and Ecuador have been highly effective at responding to the burgeoning demand, in places like the United States and Germany, among those who wish to visit rain forests and other tropical habitats. This success is manifested in increased patronage of protected areas in Costa Rica and in the status the Galápagos now enjoy as a prime ecotourism destination. In turn, demand for access to natural habitats appears, in some places, to have strengthened the case for conserving them.

Other than backpackers and a few others, people who spend the time and money required to reach a place like Monteverde or the Galápagos are pretty exacting about the quality of the goods and services they purchase. More often than not, local communities find it a challenge to supply appetizing food, clean sheets, cold beverages, and, more important than anything else, safe and reliable transportation. Consequently, the firms that provide these services, which by and large are based in capital cities or foreign countries, receive most of what ecotourists spend.

The distribution of the gains from nature-based tourism could be altered by investing in human capital and local infrastructure. Before making such an investment, though, some fundamental economic questions relating to opportunity costs would have to be faced. The returns local people are likely to enjoy as a result of investment specifically tailored to ecotourism would have to be compared to the returns they would capture if, for example, spending on general human capital—which is applicable across all sectors of the economy—were increased. Likewise, other investment options, in protected areas, for instance, would have to be evaluated as well.

Opportunity cost questions are of central concern because, as emphasized in this chapter, the environmental base for nature-based tourism is by no means secure in Costa Rica and the Galápagos. Substantial sums need to be spent so that the habitats and species that attract international visitors do not disappear, due to encroachment by farmers, ranchers, and loggers, or because of competition from newly introduced flora and fauna. Those sums greatly exceed national park services' existing financial resources.

Both Costa Rica and Ecuador are trying to raise more money by increasing the fees paid by park visitors and, in the case of the Galápagos, by increasing fees paid by the businesses that cater to visitors. In some cases, price discrimination and varying fees from site to site have yielded dividends. At the Monteverde Preserve, for example, as noted previously, different fees are charged domestic students, foreign students, international tourists visiting in groups, and others. In Ecuador, foreigners who are not students pay $80 to visit the Galápagos; meanwhile, entrance charges are modest for those mainland parks that receive few visitors. By contrast, the Costa Rican government has only just begun to move away from a policy of charging all foreigners one uniform price for admission to any park and all citizens of the country another price. Setting fees that better reflect the differences among various sorts of visitors and among various parks would allow the country to capture more revenues, while at the same time redirecting people away from places, like Manuel Antonio, where carrying capacities are probably being exceeded. Likewise, linking entrance fees to trip duration should be a central feature of pricing policy, in the Galápagos and elsewhere.

Even if countries with appealing ecotourism destinations price access efficiently, it will be rare for revenues generated to exceed the expenses of protecting natural areas. Even at Monteverde, the costs of intense maintenance and vigilant protection, which are needed to secure the reserve, exceed entrance fee revenues by a wide margin. Without donations from foreign individuals and foundations, the site would deteriorate.

The Galápagos appear to be one of those unusual places with the potential to be a net generator of funds, defined broadly to consist of foreign donations as well as users' fees. As noted in this chapter, it has long been a policy to use monies collected from visitors and tourism businesses to help pay for the management of mainland parks. If the recent decisions by the Ministry of Finance are any guide, central government officials are now casting an envious eye toward those monies as well.

But even in the archipelago, it is important not to exaggerate the size of the financial surplus that ecotourism can be expected to generate, year in and year out. Once habitat protection has been adequately provided for, not much is likely to be left over for local development initiatives or anything else.

III

Key Elements of an Integrated Strategy for Habitat Protection and Economic Progress

8

Another Approach to Habitat Conservation

Agricultural Intensification

Under the right set of circumstances, nature-based tourism and the harvesting of forest products are commercially promising and can be carried out in ways that benefit local communities and keep renewable resources intact.

For example, conditions for the sustainable production of *açaí* appear to be favorable in the Amazon estuary. The region's waterways provide easy access to Belém, where several palm heart exporters are located and where demand is strong for *açaí* fruit. In addition, output can be increased through the application of relatively simple stand improvement measures, such as seed dispersal and periodic thinning (Pollak, Mattos, and Uhl, 1995). Accordingly, quite a few smallholders in the estuary find it rewarding to raise *açaí*, which involves little or no damage to natural resources.

Another place where economic development has been environmentally benign as well as socially broad-based is the Monteverde Cloud Forest Biological Preserve. The park is undoubtedly one of the best-managed of its kind in the world, and visitors spend millions of dollars each year on meals, lodging, and souvenirs in neighboring communities; it is widely hailed as an example of what ecotourism can accomplish.

No one would argue against taking full advantage of the opportunities being offered in places like the Amazon estuary and Monteverde. However, the evidence presented in this book suggests that those opportunities are not all that common. For example, promising sites for the collection of nontimber forest commodities in the Brazilian Amazon are confined almost entirely to the estuary and floodplains (Goulding, Smith, and Mahar, 1996), which together account for just 2 percent of the entire watershed. Elsewhere, harvesting costs tend to be high, because beneficial species are not concentrated. Also, unless a road or a waterway is close at hand, transportation expenses are prohibitive.

Likewise, Monteverde is a unique place. The Alabama Quakers who first settled in the area a half-century ago applied various conservation measures, including the protection of primary forests. In addition, the TSC, which manages the reserve, has pursued habitat conservation and local development initiatives that are frequently advocated but rarely implemented.

If opportunities for sustainable tropical forest development occur mainly in special environmental and market "niches," they are best exploited by implementing small-scale projects with a narrow geographic focus. An example of such a project is one undertaken in the late 1980s and early 1990s to increase *açaí* fruit output in the Amazon estuary. With support from the Ford Foundation and the British Overseas Development Agency (ODA), a plant biologist was assigned to work with local communities on the improvement of thinning practices, the establishment of a nursery to produce better seedlings, and related pursuits. Also investigated were techniques, like pinching off flower buds, for delaying fruit production until the off-season, when prices are higher (John Rhombold, personal communication, 1996). Dozens of families, at least, benefited directly from this initiative.

In contrast with the performance of small, niche-specific efforts, large-scale undertakings have experienced serious shortcomings. During the late 1980s and early 1990s, a strong case was made for promoting activities like the extraction of nontimber products, which would conserve forests while simultaneously improving local living standards (see chapter 4). AID and other development agencies responded by funding a number of multimillion-dollar projects in the region, many of which were managed by environmental organizations. By and large, these projects' accomplishments have been disappointing, with nontimber extraction, ecotourism, and other activities having taken hold only here and there.

Heeding this experience, implementing and funding agencies have become more cautious. Efforts to promote sustainable economic activities in tropical forests and other threatened ecosystems now tend to be quite limited in scope. Moreover, it is coming to be accepted that nontimber extraction, ecotourism, logging with minimal environmental impacts, and genetic prospecting can only make a modest contribution, and that the search for effective habitat conservation measures must continue.

Agroforestry in Tropical Forest Settings

A sensible response to an understanding of the limitations of economic activities examined in chapters 4–7 of this book is to pursue a more varied array of initiatives for the purpose of saving habitats and promoting local development. One alternative that holds promise is agroforestry, which involves the interplanting of crops or forages with trees that yield construction materials, fuelwood, and other useful commodities.

Denevan et al. (1984) point out that countless generations of indigenous communities in the Amazon Basin, like their counterparts in other tropical settings, have practiced one or another form of agroforestry for millenia. Such systems are environmentally sound in that they mimic the succession of plants that emerge in a small forest clearing. At the same time, careful management is often needed, in order to maximize useful output.

Indigenous peoples are not agroforestry's only practitioners. Many of the ethnic Japanese who have migrated to the Brazilian Amazon have enjoyed considerable success in interplanting black pepper with a variety of other annual and perennial crops as well as trees (Subler and Uhl, 1990). Also, Browder (1997) reports that colonists who have settled in Rondônia are interested in agroforestry; particularly enthusiastic are those who have acquired an appreciation for the limitations of ecosystem mining, as a result of having farmed in the state for several years. Smith et al. (1995) note that there is a pronounced trend toward agroforestry throughout the region, which they say should lead to diminished pressure on forested ecosystems.

Given the system's management demands, successful agroforestry can require a substantial investment in human capital. This investment is accomplished among the ethnic Japanese through producers' associations and community networks, which channel technical and market information to their members (Subler and Uhl, 1990). Agroforestry projects supported by development agencies seek to establish the same sort of institutional infrastructure, sometimes within the public sector but more and more often in civil society.

Obviously, commercial viability also has a lot to do with favorable market conditions, as an economic study of agroforestry in the Ecuadorian Amazon makes clear (Southgate et al., 1992). In this region, it is quite common for farmers to allow laurel (*Cordia alliodora*) and other commercial timber species to grow in fields where coffee has been planted (Uquillas, Ramirez, and Seré, 1991). Research conducted during the 1980s, under the auspices of an AID project, yielded several recommendations for the refinement of this practice (Peck, 1990), as well as the technical coefficients needed for a financial analysis of the improved system.

The results of the analysis, expressed as per-hectare annual net returns, are reported in Table 8.1. Each estimated net return corresponds to a certain wage, coffee price, stumpage value, and real interest rate. The lower and upper daily wages, $3.00 and $3.50, are consistent with what unskilled laborers earn in the study area. Likewise, the two coffee prices used in the study, $0.10 and $0.15 per kilogram, are in line with the farm-level realities of recent years. By contrast, both of the stumpage values, $10 and $15 per cubic meter, are better than what resource owners typically receive. Similarly, the real interest rates they pay on loans tend to be well above the 7.5-to-10-percent range. Of course, using higher stumpage values and lower interest rates enhances profitability estimates. However, the parameters used in the study were chosen because the price of standing timber would be higher if Ecuadorian wood product markets were

Table 8.1 Annual Net Returns per Hectare for Agroforestry in the Ecuadorian Amazon, late 1980s.

	Real interest rate	
	7.5%	10.0%
Daily wage = $3.00		
lau p = $10/m³ & cof p = $0.10/kg	$48.40	$36.32
lau p = $10/m³ & cof p = $0.15/kg	164.19	154.83
lau p = $15/m³ & cof p = $0.10/kg	$61.01	$45.69
lau p = $15/m³ & cof p = $0.15/kg	176.81	164.20
Daily wage = $3.50		
lau p = $10/m³ & cof p = $0.10/kg	$15.76	$2.07
lau p = $10/m³ & cof p = $0.15/kg	131.55	120.58
lau p = $15/m³ & cof p = $0.10/kg	$28.88	$11.45
lau p = $15/m³ & cof p = $0.15/kg	144.17	129.96

SOURCE: Data from Southgate et al. (1992): 42.

more competitive and because real interest rates would be lower if financial intermediation in the country were more efficient (Southgate et al., 1992).

The financial analysis indicates that agroforestry's profitability is not affected very much by interest rate fluctuations. This is because most revenues, which are associated with coffee sales, are captured early in the agroforestry rotation. Furthermore, the system is less risky than, say, establishing and managing a tree plantation. This is because multiple commodities, rather than a single commodity, are produced (Southgate et al., 1992).

By contrast, profitability is highly sensitive to wage levels since the improved system is labor-intensive. Of even greater importance is the farm-level value of the main product of the system in eastern Ecuador, which is coffee. At the higher wage and the lower coffee price, agroforestry compares poorly with alternative land uses, including extensive cattle ranching. Another reason that the system is unprofitable is linked to the diseases that afflict coffee trees, thereby diminishing yields (Southgate et al., 1992).

The Impacts of Sectorwide Improvements in Productivity

As is the case with just about any line of commerce, agroforestry's profitability is inextricably linked to prices paid for output. Of course, the best prices are received by those producers who enjoy good access to markets. This is true whether the commodity in question is coffee or black pepper produced on a farm where interplanting is practiced, or whether it is *açaí* extracted from a riparian forest. Appreciating all this, Smith et al. (1995) emphasize that the commercial prospects for agroforestry are especially favorable where there is a passable road or a navigable waterway nearby that leads to a marketing center of one kind or another.

It is not to be expected, then, that agroforestry will take place first in remote hinterlands. The same holds true for intensified crop and livestock production, in general, which results from spending more on research and extension systems and other investments in the agricultural economy. Intensification, which is related to technological improvement, can be expected to diminish human encroachment on natural habitats. Quite often, the impacts are dramatic.

The effects that higher productivity has along agricultural frontiers are depicted in Figure 8.1.. Investment and technological change lower costs of production for all farmers and ranchers, including those located in remote areas. This enterprise-level consequence is represented in Figure 8.1 by a downward shift of marginal-cost (MC) and average-cost (AC) curves. But at the same time, commodity supplies increase, as shown in the figure. This causes the market-clearing price to fall.

It is possible for the impact of lower production to dominate the impact of lower prices. Suppose that better agricultural land-clearing techniques are developed. This would result in substantial displacement of enterprise-level cost curves along agriculture's extensive margin. Provided there are no major changes in market-level supply and price, frontier producers would raise output and would also find it profitable to deforest more land.

Much more typical, however, are investment and technological change that benefit inframarginal producers more than those in remote areas. In this case, which is illustrated in Figure 8.1, frontier producers incur losses because price declines exceed reductions in per-unit costs (AC). In the short run, migration to agriculture's extensive margin ceases, since additional deforestation is unprofitable. In time, people who already have settled on the frontier choose to abandon their farms and ranches, which are reclaimed by natural vegetation.

Agricultural intensification clearly has caused the sector's extensive margin to recede in the United States. The area used to raise crops and livestock in the country reached a maximum around 1920. (In New England, the peak occurred about 150 years ago.) Since then, productivity has skyrocketed, due to the development of hybrid corn and other technological improvements, and because of investments in roads and other infrastructure. At prevailing crop prices, which reflect low production costs in the midwestern breadbasket, production of corn, soy beans, and other commodities is uneconomical from New York to Alabama. Accordingly, much of the land in that part of the country which was farmed up to a generation ago is now covered with maturing forests.

The inverse relationship between intensified crop and livestock production and expansion of the agricultural frontier holds in other parts of the world as well. Had it not been

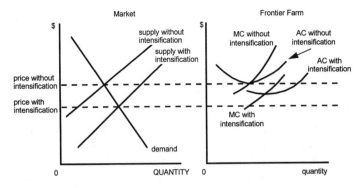

Figure 8.1 Agricultural intensification and the agricultural frontier.

for the Green Revolution, which has boosted rice and wheat yields, tropical deforestation in Asia would now be discussed mostly in the past tense. In the absence of intensification, farmers and ranchers would have had to occupy every available square meter in order to meet burgeoning food demands, resulting from population growth and higher standards of living.

Results of a regression analysis of the causes of deforestation in the Brazilian Amazon (Reis and Guzmán, 1994) could be used to bolster the claim that intensification can lead to land clearing. Indeed, those results reveal a link between increased crop output and deforestation at the county (*municipio*) level. But in no way does this disprove the fact that raising agricultural productivity tends to arrest farmers and ranchers' overall encroachment on natural habitats. Consistent with what Reis and Guzmán (1994) have found, intensification should cause cropland and pasture to expand at the expense of forests in inframarginal areas—centrally located *municipios*, in the case of the Brazilian Amazon. But at the same time, agricultural land use should be diminishing in more remote places. If more land is left undisturbed in frontier areas than is cleared in inframarginal areas, then the overall relationship between agricultural productivity increases and deforestation is negative.

That there is, in fact, a negative linkage is indicated by the findings of another regression study, one in which national-level data are used (Southgate, 1994). In that study, annual percentage growth in cropland and pasture during the middle 1980s (AGLNDGRO) is the dependent variable, and the independent variables include annual percentage growth in population, exports, and crop yields (POPGRO, EXPGRO, and YLDGRO, respectively). Also included is a dummy factor, reflecting a comparison of actual and potential land use during the early 1980s (Southgate, 1994), that indicates that land not yet cleared is poorly suited to agriculture (NO-LAND).

Using data for twenty-three countries, ordinary-least-squares estimation of the regression model yields the results that follow.

$$
\begin{aligned}
\text{AGLNDGRO} = \quad &0.463 \quad + \quad 0.249\ \text{POPGRO} \quad + \quad 0.031\ \text{EXPGRO} \\
&(0.161) \qquad (0.066) \qquad\qquad\quad (0.014) \\[1em]
&\quad - \quad 0.198\ \text{YLDGRO} \quad - \quad 0.641\ \text{NOLAND} \\
&\qquad\quad (0.033) \qquad\qquad\quad (0.205)
\end{aligned}
$$

ADJ R^2 = 0.669   SSR = 3.489 F = 12.098

All regression coefficients are statistically significant (the numbers given in parentheses are standard deviations), and each parameter estimate is consistent with what should be expected. For example, frontier expansion is positively related to population growth and increased exports. Also, in a country where natural conditions do not favor frontier expansion (i.e., where the value of NOLAND is 1 instead of 0), annual growth in cropland and pasture should be 0.641 percentage points lower than what would occur if soils lending themselves well to crop or livestock production were unoccupied.

Of special relevance to a discussion of deforestation and agricultural intensification is the negative and statistically significant coefficient of crop yield growth. (YLDGRO is a satisfactory indicator of productivity trends for the sector as a whole in this model since most deforested land is used for livestock production and, therefore, there is not much chance of lower crop yields in frontier areas having a major impact on nationwide averages.) Comparison of regression coefficients indicates that a 1 percent increase in yields can offset nearly four-fifths of the agricultural land clearing induced by a 1 percent increase in population. Alternatively, a 1 percent yield increase can compensate for the deforestation that would result from 6 percent growth in agricultural exports.

In addition to being revealed by regression analysis, the relationship between agricultural intensification and habitat conservation is clarified by the experience of two Latin American nations: Chile and Ecuador. During the 1980s, commodity demand rose dramatically in Chile, due to population growth that averaged 1.7 percent a year and an increase in agricultural exports at a 17.5 percent annual rate. Nevertheless, there was practically no geographic expansion in Chile's agricultural economy because yields were rising by 3.6 percent per annum. Meanwhile, in Ecuador, where annual population growth averaged 2.7 percent and commodity exports were rising by 11.4 percent a year, yields actually fell slightly. As a result, the country experienced the second-highest rate of agricultural land clearing in the hemisphere: 2.0 percent per annum between 1982 and 1987 (Southgate, 1994).

Commodity demand in Latin America is bound to increase for many years to come. Notwithstanding recent declines in human fertility, population growth is continuing. Also, incomes are beginning to rise, and agriculture and other sectors of the rural economy are growing because discriminatory policies, like currency overvaluation, are largely a thing of the past. Without intensification, continued deforestation is all but inevitable.

In essence, raising crop and livestock yields requires the building up of nonenvironmental wealth in the sector. Good access to markets, made possible by the articulation of local road networks, is essential, as an econometric analysis of factors contributing to agricultural land clearing in Belize makes clear. Chomitz and Gray (1995), who carried out that analysis, conclude that agricultural development can be achieved, with minimal loss of natural habitats, by improving transportation infrastructure, including secondary and feeder roads, in areas that have been settled already and where natural conditions favor crop and livestock production.

Another requirement for the sustainable intensification of agriculture is investment in the sector's science and technology base, which has been singularly deficient in Ecuador. Having declined 7.3 percent a year from 1975 through 1988, real spending on agricultural research is now less than 0.2 percent of sectoral GDP, which is low even by the standards of neighboring countries. Support for agricultural extension is similarly inadequate (Southgate and Whitaker, 1994).

An important reason why too little support has been made available for research and extension is that too much money has been spent on irrigation subsidies, which yield a poor economic return, benefit a small minority consisting of relatively affluent individuals, and result in the waste and misuse of natural resources. On average, the prices that the beneficiaries of Ecuador's public irrigation projects pay for water are around 5 percent of what it costs the public sector to deliver the input. The money needed to cover the other 95 percent, which amounts to two-fifths of the government's budget for agriculture, dwarfs the funding for the development and dissemination of improved technology for crop and livestock production. Furthermore, no beneficiary of a public irrigation project has a strong reason to adopt costly conservation measures. Accordingly, water is wasted and misused on a grand scale and land is sometimes ruined because of salinization, which is brought about by excessive irrigation and poor drainage. Furthermore, Ecuador's policy of cheap irrigation water benefits no more than 10 percent of the country's producers, who tend to be much better off than those with no connection to a public irrigation system (Southgate and Whitaker, 1994).

Since the pace at which improved crop varieties and production techniques are being developed and disseminated has been too slow in Ecuador, yields have stagnated, and commodity demand growth has had to be accommodated primarily by increasing cropland and pasture, at the expense of tropical forests and other natural habitats. Only in Surinam, a lightly populated country with very little cropland and pasture in the early 1980s, was annual percentage growth in agricultural land higher than what Ecuador experienced between 1982 and 1987 (Southgate, 1994).

Obviously, human pressure on forested hinterlands can be eased considerably by intensifying crop and livestock production. The same relationship holds true for the wood products sector. As mentioned at the end of chapter 5, future increases in global roundwood demand will be met mainly in two ways. The first is improved management in forests that are already accessible. The second response will be the establishment of tree plantations in Latin America and other parts of the developing world, and, again, in relatively accessible places. By contrast, loggers are not expected to venture far into primary forests that lack transportation infrastructure (Sohngen et al., 1997).

Intensification is not a panacea, either in agriculture or in the wood products sector. Consider the case of a small, open agricultural economy, one in which commodity prices are determined exclusively at the international level. This implies that the price impacts of technological change, which are illustrated in Figure 8.1, are bound to be negligible. If input elasticities are also very high, then improved agricultural productivity always leads to increased deforestation. Kaimowitz (1996) contends that this is an accurate description of the livestock sector's geographic expansion in Central America.

Another problem, which the econometric study by Reis and Guzmán (1994) addresses, is that agricultural intensification is often accompanied by various environmental damages. These include soil erosion and the contamination of waterways with fertilizers and pesticides. Moreover, local biodiversity is diminished insofar as intensification leads to the destruction of the last remaining patches of forests in inframarginal areas. Pagiola et al. (1997) point out that this local biodiversity can be an important buffer for agricultural systems in which one or a few species are predominant. To prevent its loss, they advocate measures like the avoidance of monoculture and the protection of forest remnants.

Notwithstanding their call to limit environmental risks, however, Pagiola et al. (1997) subscribe to the emerging consensus that intensification plays a pivotal role in arresting encroachment by farmers and ranchers on tropical forests and other natural habitats.

9

Paying for Habitat Conservation and Investing in Human and Social Capital

An important merit of increased agricultural pro-ductivity is that it facilitates other habitat protection measures. Insofar as technological change causes commodity prices to decline more than farming and ranching expenses along the extensive margins of crop and livestock production (see Figure 8.1), real estate values fall. Accordingly, it is cheaper to set aside forested hinterlands to sequester carbon, to harbor threatened species, or for any other purpose.

Of course, intensification can have precisely the opposite effect in inframarginal areas. As Reis and Guzmán's (1994) econometric analysis demonstrates and as Pagiola et al. (1997) warn, higher crop and livestock yields strengthen incentives to clear tree-covered land that is close to markets and that lends itself well to farming or ranching. Balanced against this, however, would be increased payments to keep accessible forests intact, which would be forthcoming in a society that has grown wealthier because of productivity gains in agriculture and other sectors of the economy.

Although the fate of forests within easy reach of farmers and ranchers is a legitimate concern, it is clear that intensifying crop and livestock production arrests aggregate deforestation (Southgate, 1994). As indicated in the preceding chapter, enhanced agricultural productivity has allowed for a resurgence of forests in the eastern United States, and the Green Revolution has reduced the extent of agricultural land clearing in many poor countries. Likewise, widespread intensification is absolutely essential if the visionary plans, currently being discussed, to step up habitat protection in the developing world are to succeed.

For example, Bezanson and Méndez (1995) propose the establishment of a global body under the auspices of the international Convention on Biological Diversity (CBD). It would raise money by levying charges on transnational activities that make use of the global commons, broadly defined to include oceans beyond sovereign states' exclusive economic zones, biodiversity, and other resources. Much of that money would be used to save tropical forests. Similar, though less ambitious, is a proposed scheme to use the proceeds of a tax on the international timber trade to pay for habitat conservation. For example, a 1-to-3 percent duty on tropical timber imported by the European Union would raise $30 million to $90 million per annum, without significantly distorting markets (NEI, 1989).

Better use can be made of existing mechanisms for financing park protection, such as entrance fees and the GEF. In addition, interest is growing in trust funds, which usually are created through debt-for-nature swaps arranged to benefit national park systems or specific protected areas. Among other advantages, such funds allow for the recurring costs of park management to be met and can also be used to attract complementary support from development agencies, philanthropic foundations, and other sources (Newcombe, 1995).

But putting national parks and reserves on a sounder financial footing will never be any more than a partial solution. A survey of forests in eighty countries around the world reveals that only 6 percent of them are officially protected (WWF, 1996). Even if the share in Latin America were increased to 10 percent and the money needed for effective management were raised, the vast majority of natural habitats would still be exposed to encroachment. It should not be forgotten, after all, that attempts in recent years to conserve habitats by fostering ecotourism and the sustainable harvesting of forest products grew out of an appreciation of the shortcomings of conservation strategies that were predicated exclusively on park protection.

To deal more directly with encroachment, payments reflecting the value of environmental services provided by forests could be made to agents of land use change. Available evidence of the costs of habitat destruction, summarized in chapter 2 of this book, suggests that deforestation would not be affected a great deal if farmers and ranchers internalized the downstream impacts of hydrological cycle disruptions and accelerated soil erosion. Similarly, payments that the pharmaceutical industry might offer to conserve areas where genetic raw material is collected are well below the value of newly cleared cropland or pasture along agricultural frontiers in the Western Hemisphere (see chapters 2 and 6).

By contrast, the economic disutility associated with the global climatic change induced by deforestation could well outweigh the benefits that might result from the geographic expansion of farming and ranching in the American tropics. If releasing a ton of carbon into the atmosphere creates $10 in damages and if clearing a single hectare results in 100 to 200 tons of emissions, then leaving natural vegetation undisturbed is worth $1,000 to $2,000 per hectare, over and above whatever values are attached to watershed services and biodiversity conservation. Amounts in this range greatly exceed the prices at which frontier holdings normally change hands.

It would seem that the potential exists for mutually beneficial trade between firms and individuals who gain when forests are left standing and those who benefit from land use change. To be specific, manufacturers and public utilities facing taxes or regulations on carbon emissions might find it worthwhile to offer the levels of financial inducement needed to convince landowners in places like the Amazon Basin to refrain from clearing away natural vegetation.

One alternative would be to engage in the sort of bidding exercise that the U.S. Soil Conservation Service (SCS) has used at times to identify farmer-participants in the Conservation Reserve Program (CRP). In auctions in which rights to develop tree-covered land are bought and sold, each participating landowner would state the minimum price at which he or she would agree not to remove natural vegetation from his or her holding. As a rule, such a price would comprise the difference between the present value of agricultural income and clearing costs and the present value of income derived from sustainable harvesting of timber and other forest products. On the other side of the market, bids offered by prospective buyers of development rights would reflect the cost of reducing carbon emissions in other ways, by switching to alternative energy sources, for instance.

In any market, there is a gap, reflecting bargaining and information costs, between the price paid by buyers and the price received by sellers. However, communications technology has advanced so far that the gap in the market for transferable forest development rights need not be very large. Bidding could be computerized, which would make the cost of arriving at a market equilibrium negligible. Also, plotting the boundaries of areas that would be covered by carbon sequestration agreements would be a straightforward matter, requiring the use of hand-held global positioning system (GPS) receivers that cost a few hundred dollars and that are accurate within a meter or two.

Nor would verifying compliance be much of a problem. A good indicator of the expense of monitoring is provided by a contract that the World Bank's office in Brasília recently awarded for land use assessment in five locations. Each of the locations is in the Amazon Basin and is more than 100,000 hectares in size. Since total expenditures for acquiring and analyzing satellite images and engaging in "ground-truthing" (i.e., verifying what the images appear to show) will be approximately $90,000, per-hectare costs will be less than $0.20 (Robert Schneider, personal communication, 1996). Obviously, monitoring expenses in this range would never impede a mutually beneficial exchange between landowners and those who want natural vegetation to be left undisturbed.

All the technology required for creating efficient and large-scale markets for tropical forest development rights is available. The reason why no such market exists at present is that, as a rule, carbon emissions are lightly taxed and regulated. Unless and until this situation changes, buyers of development rights will remain scarce. Carbon sequestration agreements will continue to be a novelty, involving an environmental organization here, a well-intentioned company there, either of which makes limited amounts of money available for the protection of relatively small tracts of land.

The Importance of Productivity-Enhancing Investment

Funding for habitat protection, channeled through the GEF, a CBD secretariat (if and when it is established), or some other organization, could be increased substantially and a true market could be set up for carbon sequestration services. Certainly in relation to overall capital flows, though, the impacts of all this would be minor. This is a problem because if anything less than a great deal of money is raised the impetus for deforestation will not diminish significantly.

Although the number of books and articles addressing sustainable economic development is growing large, the reasons why destruction of forests and other natural resources occurs on a grand scale in poor countries remain the subject of active debate. One explanation, which is stressed in the environmental economics literature, is that standard indicators of macroeconomic performance, which are used routinely by decision-makers in and outside government, are flawed.

As a number of economists have observed, one of the most commonly used macroeconomic indicators, net national product (NNP), does not treat physical capital evenhandedly (Solow, 1992). To arrive at NNP, deductions are made for the depreciation of equipment, buildings, and other assets made by human beings. However, depletion of oil deposits, standing timber, and other kinds of environmental wealth is entirely neglected. This means that a country can raise NNP, conventionally defined, in the short run by exhausting its natural resources, even though its long-term economic prospects might be severely impaired in the process. To illustrate the bias of standard macroeconomic measures, Repetto et al. (1989) have demonstrated that much of the Indonesian economy's apparent expansion through the middle 1980s resulted from deforestation, soil erosion, and, above all else, the depletion of fossil fuel deposits; when natural resource accounting is done, actual rates of growth turn out to have been much lower.

There have been other refinements in the conceptualization and measurement of economic progress. Although the role of natural resources is not stressed in many of the recent contributions to the literature on economic development, those contributions provide a framework for understanding why environmental degradation, generally, and habitat loss, specifically, are often excessive.

For example, Romer (1986) has examined the shortcomings of neoclassical growth models, originally developed during the 1950s. Underlying the neoclassical approach is the idea that improvements in living standards over time are linked to the accumulation of productive assets, which is reasonable enough. Moreover, capital has usually been treated as a homogeneous quantity, something that is manufactured and physical more often than not. However, the empirical models derived from theorists' work lack explanatory power. That is, attempts to estimate linkages between different nations' growth rates and their respective endowments of machinery and other manufactured capital have not met with much success. Romer (1986) contends that poor empirical results can often be traced to the neglect of less tangible assets, like human capital.

No cause of poor economic performance in Latin America and other parts of the developing world is more important than inadequate human capital formation. More often than not, providing too little support for education and health and sanitation services, which are all especially deficient in rural areas, has resulted from a fundamental misallocation of public sector resources. Over the years, enormous sums have been spent in Latin America on consumption subsidies (most of which are captured by relatively affluent households), the military, parastatal corporations, and so forth. Inevitably, education, health, and sanitation expenditures have suffered. Consequently, there are now large numbers of people in the region who do not have the skills required for more remunerative and productive employment.

Aside from poverty and lost opportunities for economic growth, there is an environmental price to be paid where education and related human capital investments are not what they should be. Environmentally sound production of agricultural commodities and timber, it must be remembered, often requires the sophisticated understanding and management of biological processes. Obviously, adoption of improved systems is discouraged if knowledge is lacking. For example, Barreto et al. (in press) identify scarce technical information as a major constraint on improved forest management in the eastern Amazon. Similarly, Browder (1997) emphasizes that accelerated training will be needed if agroforestry is to take hold in Rondônia and other parts of Brazil. At a more general level, human capital formation is essential to achieve the sustainable yield increases needed to forestall the agricultural economy's expansion onto land now covered with natural vegetation.

Environmental impacts also arise because inadequate support for education and health and sanitation services in rural areas impedes the movement of labor out of agriculture and into other sectors of the economy, typically into lines of work that involve less exploitation of natural resources. In particular, many rural people whose meager skills circumscribe their nonfarm employment possibilities find that their best option is to resort to raising crops and perhaps some livestock along hillsides, in rain forests, and in other fragile environments.

It bears repeating that mining nutrients in areas where forests are giving way to cropland and pasture is a relatively attractive option for Latin America's rural poor. By no means does this imply that the returns they derive from agricultural colonization are high. Except in advantageously located settlement projects, colonists' household incomes in the Brazilian Amazon exceed the minimum wage, which in rural areas amounts to less than $100 per month, only by a factor of two or three (FAO/UNDP/ MARA, 1992, cited in Schneider, 1995, p. 51). The returns captured by agents of deforestation in northwestern Ecuador are similarly marginal. In a survey of approximately 165 colonists carried out in 1991, Southgate et al. (1992) found that annual net cash flows in the region averaged just $25 per hectare. Of those settlers willing to name a price, the vast majority indicated that they would part with their holdings if offered just $300 per hectare.

That colonists indeed find it worth their while to migrate to remote frontiers because they lack the human capital required for more lucrative employment is revealed by surveys carried out in Ecuador. In the northeastern part of the country, average educational attainment for rural adults amounts to less than three years (Pichón, 1997). This is comparable to what Southgate et al. (1992) have found among colonist households in northwestern Ecuador.

Although it is of paramount importance, human capital scarcity is neither the only cause of rural poverty, nor the only reason why overexploitation of renewable resources is a primary feature of economic activity in the Latin American countryside. Resource degradation also results from the underdevelopment of what economists refer to as social capital, which is defined to include the rule of law, courts, and related institutions that exist to make sure that contracts and property rights are respected. Without those institutions, of course, markets function anemically or not at all.

The role of social capital has received a lot of attention in the recent literature on growth and development. For example, Olson (1996) has observed that discrepancies between poor countries' economic performance and how rich countries are faring cannot be explained only in terms of average educational levels, access to technology, the availability of machinery, and other kinds of human and physical capital. Slow growth and low living standards in places like the former Soviet Union, he argues, are caused mainly by the waste and misallocation of existing assets, which in turn have to do with the feebleness of capitalism's undergirding institutions.

Although Olson (1996) does not focus specifically on natural resources, his work furnishes an intriguing perspective on why there is not enough environmentally sustainable economic progress in the Latin American countryside. There is no denying that the resource endowments of Brazil, Venezuela, and many other countries in the Western Hemisphere are superior to what one finds in many parts of Asia and Africa, and of Europe, for that matter. However, social capital remains poorly articulated, which inevitably has environmental consequences.

The contribution that weak institutions make to tropical deforestation has been examined by Deacon (1994). He points out that poor countries' governments are often unstable, as reflected by frequent coups d'état and constitutional changes, and tend to do a poor job of enforcing property rights. Furthermore, lawlessness, plus a general lack of governmental accountability, leads to excessive habitat loss, as an econometric analysis of land clearing in 120 countries demonstrates. Deacon's (1994) results corroborate the findings of a more narrowly focused study, in which settlers in the Ecuadorian Amazon were found to respond to the government's slowness in processing applications for formal land tenure by claiming land in the time-honored way—by clearing away natural vegetation in order to exercise agricultural use rights (Southgate, Sierra, and Brown, 1991).

By contrast, strong institutions make sustainable resource management possible. For example, the success that ethnic Japanese communities have enjoyed with agroforestry in the Brazilian Amazon is related in part to the cooperative organizations they have established, which provide technical assistance, credit, market information, and other services (Subler and Uhl, 1990).

Pearce and Warford (1993) emphasize that the factors driving habitat loss and other forms of environmental degradation in poor countries are highly complex and require much more investigation. No simple explanation (e.g., that poverty causes resource depletion, or vice versa) comes close to being universally applicable. It is obvious, though, that there is a common set of causes leading both to poor economic performance (and rural poverty, in particular) and to the deterioration of renewable natural resources. Assigning specific weights to inadequate human capital and underdeveloped social capital may not be possible. Suffice it to say that the two are interrelated, and that they jointly lead to the excessive depletion of natural resources just as surely as they combine to induce the waste and misallocation of physical capital made by human hands.

The Lessons to Be Learned in El Salvador about Environmental Depletion and Conservation

Nowhere are the consequences of meager human and social capital more clearly on display than in El Salvador, which, with a national territory of only 21,041 square kilometers, is the smallest country on the American mainland.

Well into the 1980s, human fertility, which is closely linked to female illiteracy, was very high in El Salvador. But because of emigration, population growth was fairly modest, averaging just 1.8 percent per annum between the 1971 and 1992 censuses (DGEC, 1977 and 1995). But since the civil war ended, fewer people have left the country and demographic expansion has accelerated. From 1992 through 1997, the annual growth rate has been a little less than 3 percent (DGEC, 1996). Even though one out of every six or seven Salvadorans now lives in the United States, the country is more densely populated, with 250 people per square kilometer, than any other independent state in the Western Hemisphere, aside from a few small Caribbean islands (Panayotou, Faris, and Restrepo, 1997).

Economic performance in El Salvador has been positive in recent years, with structural adjustment having begun in earnest around 1990. Responding to the commercial opportunities created by freer markets, the Salvadoran economy grew by 6 percent per annum through 1995; the rate of annual expansion reached a peak of more than 7 percent immediately after the 1992 peace accords, which brought an end to a war that had lasted for more than a decade (Panayotou, Faris, and Restrepo, 1997).

However, economic growth is showing signs of flagging, the rate for 1996 amounting to only 3 percent. Some of the blame can be assigned to the exchange rate distortions resulting from remittances sent home by emigrants. These remittances exceed what the country earns from overseas sales of coffee, which is the leading national export (other than people), and, according to local bankers, have caused the Salvadoran colón to be overvalued by approximately 30 percent.

As a result, incentives to compete in international markets have been weakened and imported substitutes for locally produced commodities are artificially cheap.

But emigration, remittances, and exchange rate distortions do not fully explain why El Salvador's domestic economy is not growing much more rapidly than the country's population is. Especially in rural areas, major inefficiencies result because misguided public policies and a dearth of social capital cause markets for various factors of production to be unduly rigid. Human capital scarcity takes a severe toll as well.

One public policy that has turned out to have adverse consequences is agrarian reform. Undertaken to alleviate social ferment in the countryside, it has had the effect of keeping prime farmland tied up in relatively large, unproductive holdings. About 30 percent of the country's agricultural real estate is controlled by agrarian reform cooperatives established during the 1980s (Shaw, 1997). Nevertheless, an agricultural policy expert who works on an AID project in El Salvador confirms the widely circulated claim that less than 10 percent of the country's crop and livestock output is produced by cooperatives (Hugo Ramos, personal communication, 1997). Shaw (1997) reports that 20 percent of El Salvador's very best farmland is cooperatively owned, and that three-tenths of that real estate currently sits idle.

Left to their own devices, members of at least some cooperatives could be counted on to sell off their holdings, thereby allowing land to be transferred to firms and individuals able to use rural real estate more productively. However, this sort of exchange is preempted by various laws and regulations. Only in 1996 and 1997 have sales of cooperative land been permitted, but only to pay off debts incurred during the course of land acquisition (Shaw, 1997).

Another option would be for cooperatives to begin functioning more like regular business enterprises, like corporations, for example. But for this to happen, major improvements in social capital would be required. As they currently stand, contract and property law in the Salvadoran countryside would not provide much "assurance" to shareholders in a corporation that had evolved from a cooperative. That is, owners could not count on having their best interests respected consistently by managers and workers, given the weakness of courts and related institutions.

Currently, cooperative members appear to be responding to the social capital vacuum by refusing to invest and by selling whatever the law says is marketable (i.e., machinery but not land). When driving through the flat, fertile valleys east of San Salvador, the national capital, one is struck by the sight of primary irrigation canals leading up to the borders of extensive cooperatives, which have not organized their members to dig the secondary and tertiary canals needed to distribute water to individual fields. Likewise, farm equipment seems to have become extremely scarce on the coastal plains, where rich volcanic soils once supported a thriving cotton industry but where cooperatives now dominate and the land is severely underutilized.

A lack of social capital also appears to explain some of the inefficiencies and rigidities that characterize financial intermediation in El Salvador. The level of participation in rural credit markets is very low in the country, with only 31 percent of farmers and rural households surveyed in early 1996 reporting having taken out a loan during the preceding five years (López, 1997). Government regulation is not the problem, now that interest-rate controls are a thing of the past. Instead, limited lending in rural areas has a lot to do with poor business prospects in the agricultural sector, which in turn are related to an overvalued currency, inflexible real estate markets, and other distortions. Furthermore, restricted financial intermediation seems to reflect the fact that credit sources, both formal and informal, are not confident that borrowers can be obliged to repay. This is perhaps inevitable in a country that has gone through the sort of conflict and social dislocation that El Salvador recently experienced. (Of course, the chance that a recalcitrant borrower owns a weapon, or can easily get hold of one, has to cross a lender's mind from time to time.)

Undertaking both the policy reforms and the social capital improvements needed to make factor markets work better would reduce rural poverty and ease pressure on the natural resources on which poor people have come to depend. If land and capital markets were more efficient, real estate would end up in the hands of its highest and best users, including productive small farmers, and more investment would take place. Employment would increase, both on commercial farms and in the agribusinesses that serve them. This would benefit the rural poor directly, and would cause at least a few of them to retire from the hillside plots where they now raise corn, beans, and a few other crops.

Notwithstanding its importance, however, improving factor market efficiency is not the only condition needed for rural development that is socially broad-based as well as environmentally sustainable. Accelerated human capital formation, too, has to take place.

El Salvador's record in this area is poor. In 1990–91, health expenditures amounted to only 1.5 percent of gross domestic product (GDP). At 1.8 percent of GDP, spending on education was slightly higher, but still compared poorly with the support being provided in other countries. For example, schooling expenditures as a portion of GDP were 3.7 percent in Chile, 3.8 percent in Thailand, and 6.9 percent in Malaysia in 1990–91. Indeed, these three countries supported human capital formation better in 1960, when their standards of living were barely better than what El Salvador's are now; education expenditures as a portion of GDP at that time were 2.7 percent in Chile, 2.3 percent in Thailand, and 2.9 percent in Malaysia (Panayotou, Faris, and Restrepo, 1997).

El Salvador's poor showing in 1990–91 probably had a lot to do with its elevated military expenditures. Since then, human capital formation has claimed a greater share of government funds, and local control of schools has been increased so that teacher absenteeism can be reduced and other measures for enhancing educational quality can be applied. However, the legacy of underinvestment endures. The country's human development index, which the United Nations calculates using data on average educational attainment, life expectancy and nutritional status, and related factors, is indistinguishable from the index for Honduras, which is the poorest country in Central and South America. Only in Haiti are conditions appreciably worse (Panayotou, Faris, and Restrepo, 1997).

Education and health and sanitation services are especially deficient outside San Salvador and other major cities, and the incidence of poverty in rural areas is well above the national rate, which is 50 percent (López, 1997). An econometric analysis suggests that meager human capital does not greatly affect the income received either by small farmers or landless agricultural laborers (López, 1997). In addition, Pagiola and Dixon (1997) contend that the shortage of skills in rural areas does not greatly impede the adoption of conservation tillage and other erosion control measures. Of far greater importance are the impacts on intersectoral labor movements as well as on small farmers' locational decisions. López's (1997) econometric study reveals that the chances of finding nonagricultural work, which pays better, are positively related to educational attainment.

Rural families with limited employment prospects are particularly likely to make a go of farming on a small hillside plot. More of them are now doing exactly that in northern El Salvador, which was the scene of some of the worst fighting during the war, and which is where the steepest land in the country is found. Since the 1992 peace accords were signed, small farmers have been drifting back to the area, which had experienced a certain amount of land abandonment and forest regeneration.

Information about demographic trends in El Salvador is spotty. Especially outside cities and towns, carrying out a census in 1992, just as the civil war was ending, was difficult. In addition, several rural administrative districts, called *municipios*, have been recategorized as urban districts. Thus, census results indicating that a contraction in the rural populations of El Salvador's three northern departments occurred in the years between the last two censuses are almost certainly exaggerated (see Table 9.1). Regardless, it is clear that there has been a reversal since 1992. Although separate projections are not available for urban and rural areas (DGEC, 1996), it is reasonable to conclude that some of the population growth that has taken place in Cabañas, Chalatenango, and Morazán reflects an increase in the farming population. The increase may not be as great as rural demographic expansion has been in other parts of the country, but it is taking place nonetheless (see Table 9.1).

Resettlement of the countryside is allowing El Salvador to continue holding the distinction of being the Western Hemisphere's second most denuded country, after Haiti. Almost no tree-covered land is left for farmers to clear. Current measures of forest area include coffee plantations, where shade trees have been planted or have grown of their own accord. Not counting these plantations, which occupy 10 percent of the national territory, no more than 2 percent of El Salvador is forested (Panayotou, Faris, and Restrepo, 1997).

The Inextricable Ties Binding Habitat Protection to Economic Improvement in the Countryside

Bleak though the situation in El Salvador is in many respects, environmentalists can draw a positive lesson of great importance from the country's problems, which is that improving the economic prospects of the rural poor can contribute substantially to the conservation of Latin America's natural habitats. That is, investing in human and social capital in rural areas should enable more individuals to find more remunerative jobs, usually outside agriculture; fewer people should find themselves relegated to subsisting on a small hillside

Table 9.1   Annual Population Growth (%) in El
Salvador's Three Northern Departments

Department	1971–1992			1992–1997
	Urban	Rural	Total	Total
Cabañas	3.24	−0.54	0.33	1.64
Chalatenango	1.15	−0.31	0.13	1.66
Morazán	1.86	−0.35	0.14	1.30
Entire nation	2.97	0.80	1.77	2.91

SOURCES: Data from DGEC, 1977; DGEC, 1995; DGEC, 1996

holding, migrating to an urban slum in the hope of finding some sort of employment (informal more often than not), or settling on an agricultural frontier.

By the same token, one cannot be optimistic about economic progress and habitat conservation in the Latin American countryside if human and social capital remain scarce. In general, poverty will continue to be widespread in the countryside and the rural poor will take advantage of each and every opportunity that comes their way to make money by depleting renewable resources. Even where it is made available, full advantage will not be taken of technology required for the sustainable intensification of crop and livestock production.

Nor will it be possible to deal with the market failure of deforestation if the plight of the rural poor is not addressed more effectively. Efforts to pay for carbon sequestration and other environmental services provided by forests will be severely hampered by the costs of preventing encroachment by agricultural colonists. Likewise, initiatives to police park boundaries and to promote nature-based tourism and the sustainable development of forest resources will not withstand the human onslaught unleashed if poverty continues to grip the countryside.

Those who suppose that habitats can be protected merely by establishing nature reserves should keep in mind the geographic lengths to which poverty has driven the Salvadoran peasantry in recent decades. During the 1960s, tens of thousands pushed their way into Honduras in order to carve farms out of the neighboring country's forests; military action was required to expel them. Twenty years later, a much larger migration to the United States occurred, in spite of the efforts of the Immigration and Naturalization Service. If rural people who are desperate to support themselves and their families pay little heed to national frontiers, how much respect are they ever likely to show for park boundaries?

The economic activities examined in this book can help put a halt to deforestation. Under certain circumstances, nature-based tourism and the sustainable harvesting of timber and other forest products have a role to play. Certainly, people who have moved to agricultural frontiers should be encouraged to adopt agroforestry and other activities characterized more by resource management than by resource mining. However, natural habitats will not be safeguarded merely by fostering environmentally sound lines of work in tree-covered hinterlands. Indeed, since protection of the nonmarket environmental services that tropical forests yield is a paramount concern, perhaps it is best to reduce, rather than to enhance, the incentives to engage in all classes of economic activity, sustainable as well as depletive, in those areas. The aim, in other words, should be to make natural habitats less, not more, appealing as venues for subsistence and commerce.

Undertaking the investments required for agricultural intensification, which allows for more crops and livestock to be produced on less land, will lower the returns of frontier colonization. Complementary formation of human capital, which permits more people to find productive employment that does little or no damage to natural resources, raises the opportunity costs associated with being an agent of deforestation. Of course, strengthening capitalism's underpinning institutions and other forms of social capital stimulates all kinds of productivity-enhancing investment, which is absolutely essential for sustainable economic progress in the Latin American countryside and anywhere else in the world.

Abbreviations

AAAS   American Association for the Advancement of Science
AID   U.S. Agency for International Development
CBD   Convention on Biological Diversity
CETUR   Corporación Ecuatoriana de Turismo (Ecuadorian Tourism Corporation)
CGIAR   Consultative Group for International Agricultural Research
CI   Conservation International
CIDESA   Fundación de Capacitación e Inversión para el Desarrollo Socio-Ambiental (Foundation for Training and Investment for Socio-Environmental Development)
COFYAL   Cooperativa Forestal Yanesha Limitada (Yanesha Forestry Cooperative Limited)
CRP   Conservation Reserve Program
CSRMP   Central Selva Resource Management Project
dbh   diameter (of a tree) at breast height
DRNR   Dirección de Recursos Naturales Renovables (Venezuelan Directorate of Renewable Natural Resources)
FPCN   Fundación Peruana para la Conservación de la Naturaleza (Peruvian Foundation for the Conservation of Nature)
GDP   gross domestic product
GEF   Global Environmental Facility
GPS   global positioning system
GTZ   Gesellschaft für Technische Zusammenarbeit GmbH (German Agency for Technical Assistance)

IBGE	Instituto Brasileiro de Geografia e Estatística (Brazilian Institute of Geography and Statistics)
ICDP	Integrated Conservation and Development Project
ICT	Instituto Costarricense de Turismo (Costa Rican Tourism Institute)
IDB	Inter-American Development Bank
IFC	International Finance Corporation
IMAZON	Instituto do Homem e Meio Ambiente da Amazônia (Institute for Man and the Environment of the Amazon)
INBio	Instituto Nacional de Biodiversidad (National Biodiversity Institute of Costa Rica)
INCRA	Instituto Nacional de Colonização e Reforma Agraria (Brazil's National Institute for Colonization and Agrarian Reform)
INEFAN	Instituto Ecuatoriano Forestal, de Areas Naturales, y de Vida Silvestre (Ecuadorian Institute of Forestry, Natural Areas, and Wildlife)
INS	U.S. Immigration and Naturalization Service
IUCN	International Union for the Conservation of Nature
NNP	net national product
ODA	Overseas Development Agency
OECD	Organization for Economic Cooperation and Development
PEPP	Proyecto Especial Pichis-Palcazú (Pichis-Palcazú Special Project)
SCS	U.S. Soil Conservation Service
SPN	Servicio de Parques Nacionales (Costa Rica's National Park Service)
SPNG	Servicio del Parque Nacional Galápagos (Galápagos National Park Service)
SRP	Subsecretaría de Recursos Pesqueros (Ecuador's Subsecretariat of Fishery Resources)
SUDAM	Superintendencia de Desenvolvimento da Amazonia (Superintendency for Amazonian Development)
TNC	The Nature Conservancy
TSC	Tropical Science Center
UNESCO	United Nations Educational, Scientific, and Cultural Organization
WWF	World Wildlife Fund

References

Acosta-Solis, M. 1944. *La Tagua*. Quito: Editorial Ecuador.

Allegretti, M. 1990. "Extractive Reserves: An Alternative for Reconciling Development and Environmental Conservation in Amazonia" in A. Anderson (ed.), *Alternatives to Deforestation: Steps toward Sustainable Use of the Amazon Rain Forest*. New York: Columbia University Press.

Alston, L., G. Libecap, and R. Schneider. 1996. "The Determinants and Impact of Property Rights: Land Titles on the Brazilian Frontier," *Journal of Law, Economics, and Organization* 12:1, pp. 25–61.

Amgen Inc. 1997. *1996 Annual Report*. Thousand Oaks, Calif.

Anderson, A. 1989. "Land Use Strategies for Successful Extractive Economies." Paper read at National Wildlife Federation Symposium on Extractive Economies in Tropical Forests, Washington, D.C.

Anderson, A. and E. Ioris. 1992. "The Logic of Extraction: Resource Management and Income Generation by Extractive Populations in the Amazon Estuary" in K. Redford and C. Padoch (eds.), *Conservation of Neotropical Forests: Working from Traditional Resource Use*. New York: Columbia University Press.

Asebey, E. and J. Kempenaar. 1995. "Biodiversity Prospecting: Fulfilling the Mandate of the Biodiversity Convention," *Vanderbilt Journal of Transnational Law* 28:4, pp. 703–754.

Aylward, B. 1993. "The Economic Value of Pharmaceutical Prospecting and its Role in Biodiversity Conservation" (discussion paper 93–05), London Environmental Economics Centre.

Aylward, B., K. Allen, J. Echeverría, and J. Tosi. 1996. "Sustainable Ecotourism in Costa Rica: The Monteverde Cloud Forest Preserve," *Biodiversity Conservation* 5:3, pp. 315–343.

Baldares, M. and J. Laarman. 1990. "Derechos de Entrada a las Areas Protegidas de Costa Rica," *Ciencias Económicas* 10:1, pp. 63–82.

Banco Central del Ecuador (BCE). 1995. *Información Estadística Mensual.* Quito.

Barfod, A. 1991. "A Monographic Study of the Subfamily Phytelephantoidae," *Opera Botanica* 105, pp. 1–73.

Barreto, P., P. Amaral, E. Vidal, and C. Uhl. In press. "Costs and Benefits of Forest Management for Timber Production in Eastern Amazonia," *Forest Ecology and Management.*

Barros, A. and C. Uhl. 1995. "Logging along the Amazon River and Estuary: Patterns, Problems, and Potential," *Forest Ecology and Management* 77, pp. 87–105.

Benavides, M. and M. Pariona. 1995. "The Yanesha Forestry Cooperative and Community-Based Management in the Central Peruvian Forest" in *Proceedings of Symposium on Forestry in the Americas: Community-Based Management and Sustainability.* Madison: University of Wisconsin, Land Tenure Center.

Bermúdez, F. 1992. "Evolución del Turismo en las Areas Silvestres, Período 1982–1991," Ministerio de Recursos Naturales, Energía, y Minas, Servicio de Parques Nacionales, San José, Costa Rica.

Bermúdez, F. 1995. Unpublished national parks visitation data, from Ministerio de Recursos Naturales, Energía, y Minas, Servicio de Parques Nacionales, San José.

Bezanson, K. and R. Méndez. 1995. "Alternative Funding: Looking beyond the Nation-State," *Futures* 27:2, pp. 223–229.

Binswanger, H. 1989. "Brazilian Policies that Encourage Deforestation in the Amazon" (working paper 16), Environment Department, World Bank, Washington, D.C.

Boo, E. 1990. *Ecotourism: The Potentials and Pitfalls.* Washington, D.C.: World Wildlife Fund.

Bray, D., M. Carreón, L. Merino, and V. Santos. 1993. "On the Road to Sustainable Forestry," *Cultural Survival Quarterly* 17:1, pp. 38–41.

Bromley, R. 1981. "The Colonization of Humid Tropical Areas in Ecuador," *Singapore Journal of Tropical Geography* 2:1, pp. 15–26.

Brooke, J. 1993. "Galápagos Burden: Goats, Pigs, and Now People," *New York Times,* 30 September, p. A4.

Browder, J. 1988. "Public Policy and Deforestation in the Brazilian Amazon" in R. Repetto and M. Gillis (eds.), *Public Policies and the Misuse of Forest Resources.* Cambridge: Cambridge University Press.

Browder, J. 1992a. "Extractive Reserves and the Future of the Amazon's Rainforests: Some Cautionary Observations" in S. Counsell and T. Rice (eds.), *The Rainforest Harvest: Sustainable Strategies for Saving Tropical Forests.* London: Friends of the Earth Trust.

Browder, J. 1992b. "The Limits of Extractivism: Tropical Forest Strategies beyond Extractive Reserves," *Bioscience* 42:3, pp. 174–181.

Browder, J. 1997. "Planting the Seeds of Change in the Ashes of the Rainforest: The Rondônia Agroforestry Pilot Project," College of Architecture and Urban Studies, Virginia Polytechnic Institute and State University, Blacksburg.

Browder, J. and B. Godfrey. 1997. *Rainforest Cities: The Urban Transformation of Brazilian Amazonia.* New York: Columbia University Press.

Browder, J., E. Matricardi, and W. Abdala. 1996. "Is Sustainable Tropical Timber Production Financially Viable? A Comparative Analysis of Mahogany Silviculture among Small Farmers in the Brazilian Amazon," *Ecological Economics* 16, pp. 147–159.

Brown, K. and D. Pearce. 1994. "The Economic Value of Non-Market Benefits of Tropical Forests: Carbon Storage" in J. Weiss (ed.), *The Economics of Project Appraisal and the Environment.* London: Edward Elgar.

Burton, T. 1994. "Drug Company Looks to 'Witch Doctors' to Conjure Profits," *Wall Street Journal,* 7 July, pp. A1 and A8.

Calero-Hidalgo, R. 1992. "The Tagua Initiative in Ecuador: A Community Approach to Tropical Rain Forest Conservation and Development" in M. Plotkin and L. Famolare (eds.), *Sustainable Harvest and Marketing of Rain Forest Products.* Washington, D.C.: Island Press.

Capistrano, A. 1994. "Tropical Forest Depletion and the Changing Macroeconomy, 1967–85" in K. Brown and D. Pearce (eds.) *The Causes of Tropical Deforestation: The Economic and Statistical Analysis of Factors Giving Rise to the Loss of the Tropical Forests.* London: University College London Press.

Chase, L. 1995. "Capturing the Benefits of Ecotourism: The Economics of National Park Entrance Fees in Costa Rica" (M.S. thesis), Department of Agricultural, Resource, and Managerial Economics, Cornell University, Ithaca, N.Y.

Chomitz, K. and D. Gray. 1995. "Roads, Lands, Markets, and Deforestation: A Spatial Model of Land Use in Belize" (policy research working paper 1444), World Bank, Washington, D.C.

Chomitz, K. and K. Kumari. 1996. "The Domestic Benefits of Tropical Forests: A Critical Review Emphasizing Hydrological Functions" (policy research working paper 1601), World Bank, Washington, D.C.

Colchester, M. and L. Lohmann. 1993. *The Struggle for Land and the Fate of the Forests.* London: Zed Books.

Coles-Ritchie, M. 1996. "Analysis of Non-Timber Extractive Products from Tropical Forests: The Tagua Example in Ecuador" (M.S. thesis), Graduate School of International Studies, Bard College, Annandale-on-Hudson, N.Y.

Coomes, O. 1995. "A Century of Rain Forest Use in Western Amazonia: Lessons for Extraction-Based Conservation of Tropical Forest Resources," *Forest and Conservation History* 39:3, pp. 108–120.

Crossette, B. 1996. "Report Blames Poor Farmers for Depleting World Forests," *New York Times*, 4 August, p. 8.

Darwin, C. 1859. *On the Origin of the Species by Means of Natural Selection*. London: John Murray.

Deacon, R. 1994. "Deforestation and the Rule of Law in a Cross-Section of Countries," *Land Economics* 70:4, pp. 414–430.

de Groot, R. 1983. "Tourism and Conservation in the Galápagos Islands," *Biological Conservation* 26:4, pp. 291–300.

de Miras, C. 1994. "Las Islas Galápagos: Un Reto Económico y Tres Contradicciones Básicas," Institut Francais de Recherche Scientifique pour le Développement en Coopéracion, Quito.

Denevan, W., J. Treacy, J. Alcorn, C. Padoch, J. Denslow, and S. Flores-Paitán. 1984. "Indigenous Agroforestry in the Peruvian Amazon: Bora Indian Management of Swidden Fallows," *Interciencia* 9, pp. 346–357.

DHV Consultants BV. 1992. "Biodiversity Protection and Investment Needs for the Minimum Conservation System in Costa Rica" (report to the World Bank), Amersfoort.

Dirección General de Estadística y Censos (DGEC). 1977. *IV Censo Nacional de Población, 1971*. San Salvador: Ministerio de Economía.

Dirección General de Estadística y Censos (DGEC). 1995. *V Censo Nacional de Población, 1992*. San Salvador: Ministerio de Economía.

Dirección General de Estadística y Censos (DGEC). 1996. *Proyecciones de la Población de El Salvador, 1995–2025*. San Salvador: Ministerio de Economía.

Dixon, J. and P. Sherman. 1990. *Economics of Protected Areas: A New Look at Benefits and Costs*. Washington, D.C.: Island Press.

Dixon, R., S. Brown, R. Houghton, A. Solomon, M. Trexler, and J. Wisniewski. 1994. "Carbon Pools and Flux of Global Forest Ecosystems," *Science* 263, pp. 185–190.

Drake, S. 1991. "Local Participation in Ecotourism Projects" in T. Whelan (ed.), *Nature Tourism: Managing for the Environment*. Washington, D.C.: Island Press.

Echeverría, J., M. Hanrahan, and R. Solórzano. 1995. "Valuation of Non-Priced Amenities Provided by the Biological Resources within the Monteverde Cloud Forest Preserve, Costa Rica," *Ecological Economics* 13, pp. 43–52.

Economist. 1997a. "A Flood of Fiascos," 19 April, p. 80.

Economist. 1997b. "Galápagos: Too Many People," 10 May, p. 44.

Edwards, S. 1991. "The Demand for Galápagos Vacations: Estimation and Application to Conservation," *Coastal Management* 19:2, pp. 155–169.

Elgegren, J. 1993. "Desarrollo Sustentable y Manejo de Bosques Naturales en la Amazonía Peruana: Un Estudio Económico-Ambiental del Sistema de Manejo Forestal en Fajas en el Valle del Palcazú" (M.S. thesis), Facultad Latinoamericana de Ciencias Sociales, Quito.

Fankhauser, S. and D. Pearce. 1994. "The Social Costs of Greenhouse Gas Emissions" in *The Economics of Climate Change.* Paris: Organization for Economic Cooperation and Development.

Farnsworth, N. and D. Soejarto. 1985. "Potential Consequences of Plant Extinction in the United States on the Current and Future Availability of Prescription Drugs," *Economic Botany* 39:3, pp. 231–240.

Ferreira, V. and J. Paschoalino. 1987. "Pesquisa sobre Palmito no Instituto de Tecnología de Alimentos" in *Proceedings from the First National Conference of Researches on Palm Hearts.* Curitiba: Impresa Brasileira de Pesquisa Agropecuaria.

Fundaçao Instituto Brasileiro de Geografia e Estatística (FIBGE). 1982. *Censo Demográfico de 1980: Acre, Amazonas, Pará, Roraima, Amapá, Rondônia.* Rio de Janeiro.

Figueroa, L. 1995. "Análisis del Impacto Económico del Turismo sobre la Comunidad y sobre la Reserva Biológica Bosque Nuboso Monteverde" (report to Tropical Science Center), Servicios Corporativos Emanuel S.A., San José.

Food and Agriculture Organization of the United Nations, United Nations Development Program, and Ministerio da Agricultura e Reforma Agraria (FAO/UNDP/MARA). 1992. *Principais Indicadores Socio-Económicos dos Assentamentos de Reforma Agraria* (project report BRA-87/022). Brasília.

Goulding, M., N. Smith, and D. Mahar. 1996. *Floods of Fortune: Ecology and Economy along the Amazon.* New York: Columbia University Press.

Gradwohl, J. and R. Greenberg. 1988. *Saving the Tropical Forests.* London: Earthscan Publications.

Hamilton, L. and P. King. 1983. *Tropical Forested Watersheds: Hydrologic and Soils Response to Major Uses or Conversions.* Boulder: Westview Press.

Hartshorn, G. 1978. "Tree Falls and Tropical Forest Dynamics" in P. Tomlinson and M. Zimmermann (eds.), *Tropical Trees as Living Systems.* Cambridge: Cambridge University Press.

Hartshorn, G. 1990. "Natural Forest Management by the Yanesha Forestry Cooperative in Peruvian Amazonia" in A. Anderson (ed.), *Alternatives to Deforestation: Steps toward Sustainable Use of the Amazon Rain Forest.* New York: Columbia University Press.

Hartshorn, G., R. Simeone, and J. Tosi. 1986. "Manejo para el Rendimiento Sostenido de Bosques Naturales: Un Sinopsis del Proyecto Desarrollo del Palcazú en la Selva Central de la Amazonía Peruana," Tropical Science Center, San José.

Harvard Business School. 1992. "INBio/Merck Agreement: Pioneers in Sustainable Development" (case study NI-593-015), Boston.

Heath, J. and H. Binswanger. 1996. "Natural Resource Degradation Effects of Poverty and Population Growth Are Largely Policy

Induced: The Case of Colombia," *Environment and Development Economics* 1:1, pp. 65–84.

Hecht, S., R. Norgaard, and G. Possio. N.d. "The Economics of Cattle Ranching in Eastern Amazonia," Graduate School of Architecture and Urban Planning, University of California, Los Angeles.

Houghton, J., L. Meiro-Filho, B. Callander, N. Harris, A. Kattenberg, and K. Maskell. 1996. *Climate Change 1995: The Science of Climate Change.* Cambridge: Cambridge University Press.

Howard, A. and J. Magretta. 1995. "Surviving Success: An Interview with The Nature Conservancy's John Sawhill," *Harvard Business Review* 73:5, pp. 109–118.

Huber, R. 1996. "Case Studies Showing Costs and Benefits of Ecotourism and Cultural Heritage Protection," Sixth Caribbean Conference on Ecotourism, Point-a-Pitre, Guadeloupe.

Instituto Costarricense de Turismo (ICT). 1994. *Encuesta Aerea de Extranjeros: Epoca Alta Turística, 1994.* San José.

Instituto Costarricense de Turismo (ICT). 1995. *Anuario Estadístico de Turismo, 1994.* San José.

Instituto Nacional de Estadística y Censos (INEC). 1992. *"Análisis de los Resultados Definitivos del V Censo de Población y IV de Vivienda, Provincia de Galápagos."* Quito.

Inter-American Development Bank (IDB). 1994. "Eighth General Increase in the Financial Resources of the Inter-American Development Bank" (document AB-1704). Washington, D.C.

International Bank for Reconstruction and Development (IBRD). 1978. "Guatemala: Livestock Development Project, Loan 722-GU" (project termination report), Washington, D.C.

International Bank for Reconstruction and Development (IBRD). 1997. "Inspection Panel Finds Mixed Results in Brazilian Amazon Project" (press release) 10 April, Washington, D.C.

International Bank for Reconstruction and Development (IBRD). 1991. *The Forestry Sector: A World Bank Policy Paper.* Washington, D.C.

Ito, T. and M. Loftus. 1997. "Cutting and Dealing: Asian Loggers Target the World's Remaining Rain Forests," *U.S. News and World Report* 122:9, pp. 19–41.

Jackson, M. 1990. *Galápagos: A Natural History Guide.* Calgary: University of Calgary Press.

Kaimowitz, D. 1996. "Livestock and Deforestation in Central America in the 1980s and 1990s: A Policy Perspective," Center for International Forestry Research, Jakarta.

Katzman, M. 1977. *Cities and Frontiers in Brazil.* Cambridge: Harvard University Press.

Kramer, R., E. Mercer, and N. Sharma. 1996. "Valuing Tropical Rain Forest Protection Using the Contingent Valuation Method" in W. Adamowicz, P. Boxall, M. Luckert, W. Phillips, and W. White (eds.), *Forestry, Economics, and the Environment.* Wallingford, England: CAB International.

Laird, S. 1993. "Contracts for Biodiversity Prospecting" in W. Reid,

S. Laird, C. Meyer, R. Gámez, A. Sittenfeld, D. Jansen, M. Gollin, and C. Juma (eds.), *Biodiversity Prospecting: Using Genetic Resources for Sustainable Development*. Baltimore: World Resources Institute Publications.

Lamb, F. 1966. *Mahogany of Tropical America: Its Ecology and Management*. Ann Arbor: University of Michigan Press.

Lasser, T. 1974. *Flora de Venezuela*, 16 vols. Caracas: Instituto Botánico de la Dirección de Recursos Naturales Renovables.

Ledec, G. 1992. "The Role of Bank Credit for Cattle Raising in Financing Tropical Deforestation: An Economic Case Study" (Ph.D. dissertation), Department of Wildlife Resource Science, University of California, Berkeley.

Lemonick, M. 1995. "Can the Galápagos Survive?," *Time Magazine*, 30 October, pp. 80–82.

López, R. 1997. "Rural Poverty in El Salvador: A Quantitative Analysis" (report 16253-ES), World Bank, Washington, D.C.

MacArthur, R. and E. Wilson. 1967. *The Theory of Island Biogeography*. Princeton: Princeton University Press.

Macdonald. T. 1981. "Indigenous Responses to an Expanding Frontier: Jungle Quechua Economic Conversion to Cattle Ranching" in N. Whitten (ed.), *Cultural Transformations and Ethnicity in Modern Ecuador*. Urbana: University of Illinois Press.

Machlis, G., D. Costa, and J. Cárdenas-Salazar. 1990. "Estudio del Visitante a las Islas Galápagos," Fundación Charles Darwin, Quito.

Mahar, D. 1989. "Government Policies and Deforestation in Brazil's Amazon Region," World Bank, Washington, D.C.

Mahmood, K. 1987. "Reservoir Sedimentation: Impact, Extent, and Mitigation" (technical paper 71), World Bank, Washington, D.C.

Mattos, M. and C. Uhl. 1994. "Economic and Ecological Perspectives on Ranching in the Eastern Amazon in the 1990s," *World Development* 22:2, pp. 145–158.

McRae, M. 1997. "Is 'Good Wood' Bad for Forests?," *Science* 275, pp. 1868–1869.

Morán, E. (ed.). 1983. *The Dilemma of Amazonian Development*. Boulder: Westview Press.

Morán, E. 1989. "Adaptation and Maladaptation in Newly Settled Areas" and "Government-Directed Settlement in the 1970s: An Assessment of Transamazon Highway Colonization" in D. Schumann and W. Partridge (eds.), *The Human Ecology of Tropical Land Settlement in Latin America*. Boulder: Westview Press.

Mountfort, G. 1974. "The Need for Partnership: Tourism and Conservation," *Development Forum* 2:3, pp. 6–7.

Myers, N. 1984. *The Primary Source*. New York: Norton.

Myers, N. 1988. "Threatened Biotas: Hotspots in Tropical Forests," *Environmentalist* 8:3, pp. 1–20.

Nations, J. 1989. "La Reserva del Biósfera Maya, Petén: Estudio Técnico," Consejo Nacional de Areas Protegidas, Guatemala City.

Nelson, M. 1973. *The Development of Tropical Lands.* Baltimore: Johns Hopkins University Press.

Netherlands Economic Institute (NEI). 1989. "An Import Surcharge on the Import of Tropical Timber in the European Community: An Evaluation," Rotterdam.

Newcombe, K. 1995. "Financing Innovations and Instruments: Contribution of the Investment Portfolio of the Pilot Phase of the Global Environmental Facility" in McNeely, J. (ed.), *Biodiversity Conservation in the Asia and Pacific Region.* Manila: Asian Development Bank.

Oficina Costarricense de Implementación Conjunta (OCIC). 1996. *Costa Rica-Norway Reforestation and Forest Conservation AIJ Pilot Project.* San José.

Olivera, A. 1995. "Forestry Project of the Indigenous Chiquitano Communities of Lomerío" in *Proceedings of Symposium on Forestry in the Americas: Community-Based Management and Sustainability.* Madison: University of Wisconsin, Land Tenure Center.

Olson, M. 1996. "Big Bills Left on the Sidewalk: Why Some Nations Are Rich and Others Poor," *Journal of Economic Perspectives* 10:2, pp. 3–24.

Ottaway, M. 1995. "Pick of the Bunch: Costa Rica is Central America at its Very Best" *Sunday Times* (London), 19 November, pp. 5.1–5.2.

Pagiola, S. and J. Dixon. 1997. "Land Degradation Problems in El Salvador" (report 16253-ES), World Bank, Washington, D.C.

Pagiola, S. and J. Kellenberg with L. Vidaeus and J. Srivastava. 1997. "Mainstreaming Biodiversity in Agricultural Development: Toward Good Practice" (environment paper 15), World Bank, Washington.

Panayotou, T., R. Faris, and C. Restrepo. 1997. *El Desafío Salvadoreño: De la Paz al Desarrollo Sostenible.* San Salvador: Fundación Salvadoreña para el Desarrollo Económico y Social.

Pearce, D. 1996. "Global Environmental Value and the Tropical Forests: Demonstration and Capture" in W. Adamowicz, P. Boxall, M. Luckert, W. Phillips, and W. White (eds.), *Forestry, Economics, and the Environment.* Wallingford, England: CAB International.

Pearce, D. and K. Brown. 1994. "Saving the World's Tropical Forests" in K. Brown and D. Pearce (eds.) *The Causes of Tropical Deforestation: The Economic and Statistical Analysis of Factors Giving Rise to the Loss of the Tropical Forests.* London: University College-London Press.

Pearce, D. and S. Puroshothaman. 1995. "The Economic Value of Plant-Based Pharmaceuticals" in T. Swanson (ed.), *Intellectural Property Rights and Biodiversity Conservation: An Interdisciplinary Analysis of the Values of Medicinal Plants.* Cambridge: Cambridge University Press.

Pearce, D. and J. Warford. 1993. *World without End: Economics,*

Environment, and Sustainable Development. New York: Oxford University Press.

Peck, R. 1990. "Promoting Agroforestry Practices among Small Producers: The Case of the Coca Agroforestry Project in Amazonian Ecuador" in A. Anderson (ed.), *Alternatives to Deforestation: Steps toward Sustainable Use of the Amazon Rain Forest.* New York: Columbia University Press.

Peters, C. 1990. "Population Ecology and Management of Forest Fruit Trees in Peruvian Amazonia" in A. Anderson (ed.), *Alternatives to Deforestation: Steps toward Sustainable Use of the Amazon Rain Forest.* New York: Columbia University Press.

Peters, C., A. Gentry, and R. Mendelsohn. 1989. "Valuation of an Amazon Rainforest," *Nature* 339, pp. 655–656.

Pichón, F. 1997. "Colonist Land Allocation Decisions: Land Use and Deforestation in the Ecuadorian Amazon Frontier," *Economic Development and Cultural Change,* 45:4, pp. 707–744.

Plotkin, M. 1994. *Tales of a Shaman's Apprentice: An Ethnobotanist Searches for New Medicines in the Amazonian Rainforest.* New York: Viking Penguin.

Pollak, H., M. Mattos, and C. Uhl. 1995. "A Profile of Palm Heart Extraction in the Amazon Estuary" *Human Ecology,* 23:3, pp. 357–385.

Principe, P. 1989. "The Economic Significance of Plants and their Constituents as Drugs" in H. Wagner, H. Hikino, and N. Farnsworth (eds.), *Economic and Medicinal Plant Research,* vol. 3. London: Academic Press.

Redford, K. 1992. "The Empty Forest," *Bioscience* 42:6, pp. 412–422.

Reid, W., S. Laird, R. Gámez, A. Sittenfeld, D. Jansen, M. Gollin, and C. Juma. 1993. "A New Lease on Life" in W. Reid, S. Laird, C. Meyer, R. Gámez, A. Sittenfeld, D. Jansen, M. Gollin, and C. Juma (eds.), *Biodiversity Prospecting: Using Genetic Resources for Sustainable Development.* Baltimore: World Resources Institute Publications.

Reis, E. and R. Guzmán. 1994. "An Econometric Model of Amazon Deforestation" in K. Brown and D. Pearce (eds.) *The Causes of Tropical Deforestation: The Economic and Statistical Analysis of Factors Giving Rise to the Loss of the Tropical Forests.* London: University College London Press.

Repetto, R. and M. Gillis (eds.). 1988. *Public Policies and the Misuse of Forest Resources.* Cambridge: Cambridge University Press.

Repetto, R., W. Magrath, M. Wells, C. Beer, and F. Rossini. 1989. *Wasting Assets: Natural Resources in National Income Accounts.* Washington, D.C.: World Resources Institute.

Richards, E. 1991. "The Forest *Ejidos* of Southeast Mexico: A Case Study of Community-Based Sustained Yield Management," *Commonwealth Forestry Review* 70:4, pp. 290–311.

Robinson, L. 1997. "Latin Robin Hood: Brazil's Squatter Movement

Pits Church against State," *U.S. News and World Report* 122: 24, pp. 30–32.

Romanoff, S. 1981. "Análisis de las Condiciones Socioeconómicas para el Desarrollo Integral de la Amazonía Boliviana" (consulting report), Organization of American States, Washington, D.C.

Romer, P. 1986. "Increasing Returns and Long-Run Growth" *Journal of Political Economy* 94:5, pp. 1002–1037.

Rovinski, Y. 1991. "Private Reserves, Parks, and Ecotourism in Costa Rica" in T. Whelan (ed.), *Nature Tourism: Managing for the Environment.* Washington, D.C.: Island Press.

Ruitenbeek, H. 1989. "Social Cost-Benefit Analysis of the Korup Project, Cameroon" (consulting report), Worldwide Fund for Nature, London.

Salafsky, N., B. Dugelby, and J. Terborgh. 1992. "Can Extractive Reserves Save the Rainforest?," Duke University Center for Tropical Conservation, Durham, N.C.

Salati, E. and P. Bose. 1984. "Amazon Basin: A System in Equilibrium," *Science* 225, pp. 129–138.

Scartezini, A. 1985. *Segredos de Medici.* São Paulo: Editora Marco Zero.

Schneider, R. 1992. "Brazil: An Analysis of Environmental Problems in the Amazon" (report 9104-Br), World Bank, Washington, D.C.

Schneider, R. 1995. "Government and the Economy on the Amazon Frontier" (environment paper 11), World Bank, Washington, D.C.

Schwartzman, S. 1989. "Extractive Reserves: The Rubber Tappers' Strategy for Sustainable Use of the Amazon Rain Forest" in J. Browder (ed.), *Fragile Lands of Latin America: Strategies for Sustainable Development.* Boulder, Colorado: Westview Press.

Seitz, F. 1996. "A Major Deception on Global Warming," *Wall Street Journal,* 12 June, p. A16.

Shaw, C. 1997. "Rural Land Markets" (report 16253-ES), World Bank, Washington, D.C.

Simeone, R. 1990. "Land Use Planning and Forestry-Based Economy: The Case of the Amuesha Forestry Cooperative," *Tebiwa: The Journal of the Idaho Museum of Natural History* 24, pp. 7–12.

Simpson, D. and R. Sedjo. 1996. "Paying for the Conservation of Endangered Ecosystems: A Comparison of Direct and Indirect Approaches," *Environment and Development Economics* 1:2, pp. 241–257.

Simpson, D., R. Sedjo, and J. Reid. 1996. "Valuing Biodiversity: An Application to Genetic Prospecting," *Journal of Political Economy* 104:1, pp. 163–185.

Smith, N. 1981. "Colonization Lessons from a Tropical Forests," *Science* 214, pp. 755–761.

Smith, N., T. Fik, P. Alvim, I. Falesi, and E. Serrao. 1995. "Agroforestry Developments and Potential in the Brazilian Amazon," *Land Degradation and Rehabilitation* 6, pp. 251–263.

Smith, R. 1982. "Las Comunidades Nativas y el Mito del Grán Vacio Económico: Un Análisis de Planificación para el Desarrollo en el Proyecto Especial Pichis Palcazú" (consulting report), U.S. Agency for International Development, Lima.

Snook, L. 1993. "Stand Dynamics of Mahogany (*Swietenia Macrophylla* King) after Fire and Hurricanes in the Tropical Forests of the Yucatan Peninsula, Mexico." (Ph.D. dissertation), School of Forestry, Yale University, New Haven.

Sohngen, B., R. Mendelsohn, R. Sedjo, and K. Lyon. 1997. "An Analysis of Global Timber Markets" (discussion paper 97–37), Resources for the Future, Washington.

Solow, R. 1992. "An Almost Practical Step toward Sustainability." Lecture presented at Resources for the Future, Washington, D.C.

Southgate, D. 1994. "Tropical Deforestation and Agricultural Development in Latin America" in K. Brown and D. Pearce (eds.), *The Causes of Tropical Deforestation: The Economic and Statistical Analysis of Factors Giving Rise to the Loss of Tropical Forests*. London: University College, London Press.

Southgate, D. and H. Clark. 1993. "Can Conservation Projects Save Biodiversity in South America?," *Ambio* 22:2–3, pp. 163–166.

Southgate, D., M. Coles-Ritchie, and P. Salazar-Canelos. 1996. "Can Tropical Forests Be Saved by Harvesting Non-Timber Products? A Case Study for Ecuador" in W. Adamowicz, P. Boxall, M. Luckert, W. Phillips, and W. White (eds.), *Forestry, Economics, and the Environment*. Wallingford, England: CAB International.

Southgate, D. and J. Elgegren. 1995. "Development of Tropical Timber Resources by Local Communities: A Case Study from the Peruvian Amazon," *Commonwealth Forestry Review* 74:2, pp. 142–146.

Southgate, D., with M. Hanrahan, M. Bonifaz, M. Camacho, M. Carey, and L. Chase. 1992. "The Economics of Agricultural Land Clearing in Northwestern Ecuador," Instituto de Estrategias Agropecuarias, Quito.

Southgate, D., R. Sierra, and L. Brown. 1991. "A Statistical Analysis of the Causes of Deforestation in Eastern Ecuador," *World Development* 19:9, pp. 1145–1151.

Southgate, D. and M. Whitaker. 1994. *Economic Progress and the Environment: One Developing Country's Policy Crisis*. New York: Oxford University Press.

Spruce, R. 1970. *Notes of a Botanist on the Amazon and Andes, Volume II*. London: Johnson Reprint Corp.

Stone, S. 1996. "Economic Trends in the Timber Industry of the Brazilian Amazon: Evidence from Paragominas" (CREED working paper 6), International Institute for Environment and Development, London.

Subler, S. and C. Uhl. 1990. "Japanese Agroforestry in Amazonia: A Case Study in Tomé-Açu, Brazil" in A. Anderson (ed.), *Al-*

ternatives to Deforestation: Steps toward Sustainable Use of the Amazon Rain Forest. New York: Columbia University Press.

Swanson, T. 1995. "Diversity and Sustainability: Evolution, Information, and Institutions" in T. Swanson (ed.), *Intellectual Property Rights and Biodiversity Conservation: An Interdisciplinary Analysis of the Values of Medicinal Plants*. Cambridge: Cambridge University Press.

Tobias, D. and R. Mendelsohn. 1991. "Valuing Ecotourism in a Tropical Rain Forest Preserve," *Ambio* 20:2, pp. 91–93.

Tosi, J. 1986. "Natural Forest Management for the Sustained Yield of Forest Products," Tropical Science Center, San José.

Uhl, C., P. Barreto, A. Veríssimo, E. Vidal, P. Amaral, A. Barros, C. Souza, J. Johns, and J. Gerwing. 1997. "Natural Resource Management in the Brazilian Amazon: An Integrated Approach," *Bioscience* 47:3, pp. 160–168.

Uhl, C. and J. Kauffman. 1990. "Deforestation, Fire Susceptibility, and the Potential Response of Tree Species to Fire in the Eastern Amazon," *Ecology* 71:2, pp. 437–449.

Uhl, C., A. Veríssimo, M. Mattos, Z. Brandino, and I. Vieira. 1991. "Social, Economic, and Ecological Consequences of Selective Logging in an Amazon Frontier: The Case of Tailandia," *Forest Ecology and Management* 46, pp. 243–273.

Umaña, A. and K. Brandon. 1992. "Inventing Institutions for Conservation: Lessons from Costa Rica" in S. Annis (ed.), *Poverty, Natural Resources, and Public Policy in Central America*. New Brunswick, N.J.: Transaction Publishers.

Uquillas, J., A. Ramirez, and C. Seré. 1991. "Are Modern Agroforestry Practices Economically Viable? A Case Study in the Ecuadorian Amazon," Workshop on the Economics of Agroforestry Systems, the Nitrogen Tree Fixing Association, Honolulu.

Vásquez, R. and A. Gentry. 1989. "Use and Misuse of Forest-Harvested Fruits in the Iquitos Area," *Conservation Biology* 3: 4, pp. 350–361.

Veríssimo, A., P. Barreto, M. Mattos, R. Tarifa, and C. Uhl. 1992. "Logging Impacts and Prospects for Sustainable Forest Management in an Old Amazonian Frontier: The Case of Paragominas," *Forest Ecology and Management* 55, pp. 169–199.

Veríssimo, A., P. Barreto, R. Tarifa, and C. Uhl. 1995. "Extraction of a High-Value Natural Resource in Amazonia: The Case of Mahogany," *Forest Ecology and Management* 72, pp. 39–60.

Vogel, J. 1994. *Genes for Sale: Privatization as a Conservation Policy*. New York: Oxford University Press.

von Thünen, J. 1866. *The Isolated State*. New York: Pergamin Press.

Weiner, J. 1994. "The Handy-Dandy Evolution Prover," *New York Times Magazine*, 8 May, pp. 40–41.

Wells, M. and K. Brandon. 1993. *People and Parks: Linking Protected Area Management with Local Communities*. Washington, D.C.: World Bank.

Whelan, T. 1991. "Ecotourism and its Role in Sustainable Development" in T. Whelan (ed.), *Nature Tourism: Managing for the Environment*. Washington, D.C.: Island Press.

Williams, R. 1986. *Export Agriculture and the Crisis in Central America*. Chapel Hill: University of North Carolina Press.

Wilson, E. (ed.). 1988. *Biodiversity*. Washington, D.C.: National Academy Press.

World Resources Institute (WRI). 1990. *World Resources 1990–91*. Washington, D.C.

World Resources Institute (WRI). 1996. *World Resources 1996–97*. Washington, D.C.

World Wildlife Fund (WWF). 1996. "First-of-Its-Kind World Forest Map Reveals 94 Percent of Earth's Forests Have No Formal Protection" (press release), Washington, D.C.

Zador, M. 1994. "Galápagos Marine Resources Reserve: A Pre-Investment Analysis for the Parks in Peril Program," The Nature Conservancy, Washington, D.C.

Index